Calculus FOR BEGINNERS

The Ultimate Step by Step Guide to Acing Calculus

By

Reza Nazari

Copyright © 2024

Effortless Math Education Inc.

All rights reserved. No part of this publication may be reproduced, stored in a retrieval system, or transmitted in any form or by any means, electronic, mechanical, photocopying, recording, scanning, or otherwise, except as permitted under Section 107 or 108 of the 1976 United States Copyright Act, without permission of the author.

All inquiries should be addressed to:
info@effortlessMath.com
www.EffortlessMath.com

ISBN: 978-1-63719-567-3

Published by: **Effortless Math Education Inc.**

for Online Math Practice Visit www.EffortlessMath.com

Welcome to
Calculus Prep

2024

Thank you for choosing Effortless Math for your Calculus preparation and congratulations on making the decision to take the Calculus course! It's a remarkable move you are taking, one that shouldn't be diminished in any capacity.

That's why you need to use every tool possible to ensure you succeed on the final exam with the highest possible score, and this extensive study guide is one such tool.

Calculus for Beginners is designed to be comprehensive and cover all the topics that are typically covered in a *calculus* course. It provides clear explanations and examples of the concepts and includes practice problems and quizzes to test your understanding of the material. The textbook also provides step-by-step solutions to the problems, so you can check your work and understand how to solve similar problems on your own.

Additionally, this book is written in a user-friendly way, making it easy to follow and understand even if you have struggled with math in the past. It also includes a variety of visual aids such as diagrams, graphs, and charts to help you better understand the concepts.

Calculus for Beginners is flexible and can be used to supplement a traditional classroom setting, or as a standalone resource for self-study. With the help of this comprehensive textbook, you will have the necessary foundation to master the material and succeed in the Calculus course.

Effortless Math's Calculus Online Center

Effortless Math Online *Calculus* Center offers a complete study program, including the following:

- ✓ Step-by-step instructions on how to prepare for the *Calculus* test

- ✓ Numerous *Calculus* worksheets to help you measure your math skills

- ✓ Complete list of *Calculus* formulas

- ✓ Video lessons for all *Calculus* topics

- ✓ Full-length *Calculus* practice tests

- ✓ And much more…

No Registration Required

Visit **EffortlessMath.com/calculus** to find your online calculus resources.

How to Use This Book Effectively?

Look no further when you need a study guide to improve your math skills to succeed on the *Calculus* course. Each chapter of this comprehensive guide to the *Calculus* will provide you with the knowledge, tools, and understanding needed for every topic covered on the course.

It's very important that you understand each topic before moving onto another one, as that's the way to guarantee your success. Each chapter provides you with examples and a step-by-step guide of every concept to better understand the content that will be on the course. To get the best possible results from this book:

➢ **Begin studying long before your final exam date.** This provides you ample time to learn the different math concepts. The earlier you begin studying for the test, the sharper your skills will be. Do not procrastinate! Provide yourself with plenty of time to learn the concepts and feel comfortable that you understand them when your test date arrives.

➢ **Practice consistently.** Study Calculus concepts at least 30 to 40 minutes a day. Remember, slow and steady wins the race, which can be applied to preparing for the Calculus test. Instead of cramming to tackle everything at once, be patient and learn the math topics in short bursts.

➢ Whenever you get a math problem wrong, **mark it off, and review it later** to make sure you understand the concept.

➢ Start each session by **looking over the previous material.**

➢ Once you've reviewed the book's lessons, **take a practice test at the back of the book** to gauge your level of readiness. Then, review your results. Read detailed answers and solutions for each question you missed.

➢ **Take another practice test** to get an idea of how ready you are to take the actual exam. Taking the practice tests will give you the confidence you need on test day. Simulate the Calculus testing environment by sitting in a quiet room free from distraction. Make sure to clock yourself with a timer.

Looking for more?

Visit EffortlessMath.com/Calculus to find hundreds of Calculus worksheets, video tutorials, practice tests, Calculus formulas, and much more.

Or scan this QR code.

No Registration Required.

What is Calculus?

- Calculus is a branch of mathematics that deals with the study of functions, limits, derivatives, integrals, and infinite series. It is a powerful tool for solving problems in various fields such as physics, engineering, economics, and statistics.

- At its core, calculus is concerned with the analysis of rates of change and the accumulation of small changes over time.

- The fundamental concept of calculus is the limit, which describes the behavior of a function as its input approaches a certain value. Limits are used to define derivatives, which measure the rate of change of a function at a given point. Derivatives are used to solve optimization problems, such as finding the maximum or minimum value of a function.

- Integrals are the reverse operation of derivatives and are used to calculate the total accumulation of small changes over a given interval. They are used to solve problems involving area, volume, and probability. Infinite series are also an important part of calculus, representing the sum of an infinite number of terms.

- Calculus is divided into two main branches: differential calculus and integral calculus. Differential calculus deals with the study of derivatives and their applications, while integral calculus deals with the study of integrals and their applications. These two branches are closely related and are used together to solve a wide range of mathematical problems.

- Overall, calculus is a powerful and versatile tool that has revolutionized the way we solve problems in mathematics and beyond. Its applications are vast and continue to grow in importance and relevance.

Importance of Calculus

- Calculus allows us to comprehend and model changes in quantities. It provides a precise framework to study rates of change and how quantities interact with each other, enabling us to analyze dynamic systems and phenomena.

- Many fundamental laws of physics, such as Newton's laws of motion and laws of thermodynamics, are expressed using calculus. It serves as a vital tool for formulating and solving equations that describe the behavior of physical systems.

- Calculus aids in optimization problems, where the goal is to maximize or minimize a particular quantity. This has implications in fields like economics, engineering, and logistics, enabling us to make informed decisions and find the best possible outcomes.

- Calculus helps in understanding forces, and energy. It is used in areas such as mechanics, electromagnetism, fluid dynamics, and structural analysis.

- Calculus plays a crucial role in economic models, optimizing production and consumption, analyzing market behavior, and determining optimal pricing and investment strategies. It enables economists and financial analysts to make informed predictions and decisions.

- Calculus underlies statistical methods, regression analysis, and data modeling. It enables us to estimate trends, fit curves to data, and make predictions based on statistical models.

Calculus in Real Life

- Calculus is everywhere in engineering. It's used to design bridges, predict how fluids move, and analyze the performance of machines. When we use a machine or cross a bridge, the principles of calculus are at work to make sure everything functions correctly.

- In economics, calculus helps predict how markets will act. It's used to find the best ways to produce goods, how to price them, and how to understand buying behaviors. Businesses and governments rely on these predictions to make decisions.

- Medicine uses calculus in various ways. It helps decide how much medicine a patient needs and predicts how diseases spread. When doctors make decisions about treatments or public health, calculus often plays a role.

- In computers, calculus helps create algorithms for machine learning, design graphics for games, and predict how systems will behave. It's fundamental in processing and analyzing the vast amounts of data we use today.

- Calculus also helps us understand the environment. It can predict things like river flows or changes in weather patterns.

- In astronomy, calculus predicts planetary motion, evaluates star brightness, and calculates orbits.

- Moreover, in sports, coaches and analysts use calculus to study player performance and plan strategies.

Contents

Chapter 1: Functions — 1

- Function Notation 2
- Adding and Subtracting Functions 3
- Multiplying and Dividing Functions 4
- Composition of Functions 5
- Writing Functions 6
- Graphing Functions 7
- Parent Functions 8
- Function Inverses 9
- Inverse Variation 10
- Domain and Range of functions 11
- Piecewise Function 12
- Positive, Negative, Increasing and Decreasing Functions on Intervals 13
- Chapter 1: Practices 14
- Chapter 1: Answers 18

Chapter 2: Advanced Functions — 21

- Exponential Function 22
- Linear, Quadratic and Exponential Models 23
- Linear vs Exponential Growth 24
- Logarithms 25
- Properties of logarithms 26
- Natural Logarithms 27
- Sine, Cosine, and Tangent 28
- Reciprocal Functions: Cosecant, Secant, and Cotangent 29
- Domain and Range of Trigonometric Functions 30
- Trigonometric Function Values for Key Angles 31
- The Unit Circle 32
- Additional trigonometric reminders 33
- Periodic properties of trigonometric functions 34
- Floor and Ceiling Functions 35
- Chapter 2: Practices 36
- Chapter 2: Answers 38

Chapter 3: Sequences and Series — 39

- Arithmetic Sequences 40
- Geometric Sequences 41
- Sigma Notation (Summation Notation) 42
- Arithmetic Series 43
- Geometric Series 44
- Binomial Theorem 45
- Pascal's Triangle 46

Chapter 3

Alternate Series	47
Chapter 3: Practices	48
Chapter 3: Answers	52

Chapter: Limit and Continuity — 53

Limit Introduction	54
Neighborhood	55
Estimating Limits from Tables	56
Functions with Undefined Limits (from table)	57
Functions with Undefined Limits (from graphs)	58
One Sided Limits	59
Limit at Infinity	60
Continuity at a Point	61
Continuity over an Interval	62
Removing Discontinuity	63
Direct Substitution	64
Limit Laws	65
Limit Laws Combinations	66
The Squeeze Theorem	67
Indeterminate and Undefined	68
Infinity cases	69
Trigonometric Limits	70
Rationalizing Trigonometric Functions	71
Algebraic Manipulation	72
Redefining function's value	73
Rationalizing Infinite Limits	74
Chapter 4: Practices	75
Chapter 4: Answers	78

Chapter: Derivative — 79

Derivative introduction	80
Average and instantaneous rates of change	81
The derivative of a function	82
Derivative of trigonometric functions	83
Power rule	84
Product rule	85
Quotient rule	86
Chain rule	87
Power rule combined with other derivative rules	88
Derivative of radicals	89
Derivative of logarithms and exponential functions	90
L'Hôpital	91
Differentiability	92
Second derivatives: Minimum vs. Maximum	93
Curve sketching using derivatives	94
Differentiating Inverse Functions	95

Contents

5
- Optimization problems 96
- Implicit differentiation 97
- Related rates 98
- Chapter 5: Practices 99
- Chapter 5: Answers 102

Chapter: Integrals — 105

6
- What is Integral? 106
- Applications of Integrals 107
- Exponential Growth and Decay 108
- The Anti-Derivative 109
- Riemann Sums 110
- Rules of Integration 111
- Power Rule 112
- Fundamental Theorem of Calculus 113
- Trigonometric Integrals 114
- Substitution Rule 115
- Integration by Parts 116
- Integral of Radicals 117
- Exponential and Logarithmic Integrals 118
- Improper Integrals 119
- Chapter 6 Practices 120
- Chapter 6: Answers 122

Chapter: Differential Equations — 123

7
- Introduction and applications 124
- Classification of Differential Equations 125
- First-Order Ordinary Differential Equations 126
- Linear Differential Equations 127
- Separable Differential Equations 128
- Slope Fields 129
- Euler's Method for Numerical Solutions 130
- Simple Growth and Decay 131
- Population Models 132
- Chapter 7: Practices 133
- Chapter 7: Answers 136

Chapter: Analytic Geometry — 137

8
- Ellipses, parabolas, and hyperbolas 138
- Polar Coordinates 139
- Converting Between Polar and Rectangular Coordinates ... 140
- Graphing Polar Equations 141
- Applications of Polar Coordinates 142
- Chapter 8: Practices 143
- Chapter 8: Answers 144

Chapter 9: Complex Numbers — 145

- Complex Numbers addition and subtraction 146
- Multiplying and Dividing Complex Numbers 147
- Rationalizing Imaginary Denominators.. 148
- Chapter 9: Practices... 149
- Chapter 9: Answers 150

Time to Test... 151
Calculus Practice Test 1 152
Calculus Practice Test 2 168
Calculus Practice Tests Answer Keys... 186
Calculus Practice Tests Answers and Explanations 187

CHAPTER 1
Functions

Math topics that you'll learn in this chapter:

- ☑ Function Notation
- ☑ Adding and Subtracting Functions
- ☑ Multiplying and Dividing Functions
- ☑ Composition of Functions
- ☑ Writing Functions
- ☑ Graphing Functions
- ☑ Parent Functions
- ☑ Function Inverses
- ☑ Inverse Variation
- ☑ Domain and Range of Functions
- ☑ Piecewise Function
- ☑ Positive, Negative, Increasing and Decreasing Functions

Function Notation

- Functions in math assign distinct output values to input values. For instance, when monitoring a baby's monthly weight, "distinct" means each month gets a single weight entry, even if the baby's weight is the same for two different months. This highlights that each input corresponds to a unique output.

- Functions relate inputs to outputs, model behaviors, simplify tasks, guide predictions, and structure math and code.

- Function notation is the way a function is written. It is meant to be a precise way of giving information about the function without a rather lengthy written explanation.

- The most popular function notation is $f(x)$ which is read as "f of x". Any letter can name a function. For example: $g(x)$, $h(x)$, etc.

- To evaluate a function, plug in the input (the given value or expression) for the function's variable (or place holder, x).

Examples:

Example 1. Evaluate: $f(x) = x + 6$, find $f(2)$.

Solution: Substitute x with 2:

Then: $f(x) = x + 6 \Rightarrow f(2) = 2 + 6 \Rightarrow f(2) = 8$.

Example 2. Evaluate: $w(x) = 3x - 1$, find $w(4)$.

Solution: Substitute x with 4:

Then: $w(x) = 3x - 1 \Rightarrow w(4) = 3(4) - 1 = 12 - 1 = 11$.

Example 3. Evaluate: $f(x) = 2x^2 + 4$, find $f(-1)$.

Solution: Substitute x with -1:

Then: $f(x) = 2x^2 + 4 \Rightarrow f(-1) = 2(-1)^2 + 4 \Rightarrow f(-1) = 2 + 4 = 6$.

Example 4. Evaluate: $h(x) = 4x^2 - 9$, find $h(2a)$.

Solution: Substitute x with $2a$: We know that $h(x) = 4x^2 - 9$, then:

$h(2a) = 4(2a)^2 - 9 \Rightarrow h(2a) = 4(4a^2) - 9 \Rightarrow h(2a) = 16a^2 - 9$.

Adding and Subtracting Functions

- Just like we can add and subtract numbers and expressions, we can add or subtract two functions and simplify or evaluate them. The result is a new function.

- For two functions $f(x)$ and $g(x)$, we can create two new functions:

$$(f+g)(x) = f(x) + g(x) \text{ and } (f-g)(x) = f(x) - g(x)$$

Examples:

Example 1. If $g(x) = 2x - 2$, $f(x) = x + 1$. Find: $(g + f)(x)$.

Solution: We know that: $(g + f)(x) = g(x) + f(x)$.

Then: $(g + f)(x) = (2x - 2) + (x + 1) = 2x - 2 + x + 1 = 3x - 1$.

Example 2. If $f(x) = 4x - 3$, $g(x) = 2x - 4$. Find: $(f - g)(x)$.

Solution: Considering that: $(f - g)(x) = f(x) - g(x)$.

Then: $(f - g)(x) = (4x - 3) - (2x - 4) = 4x - 3 - 2x + 4 = 2x + 1$.

Example 3. If $g(x) = x^2 + 2$, and $f(x) = x + 5$. Find: $(g + f)(x)$.

Solution: According to the: $(g + f)(x) = g(x) + f(x)$.

Then: $(g + f)(x) = (x^2 + 2) + (x + 5) = x^2 + x + 7$.

Example 4. If $f(x) = 5x^2 - 3$, and $g(x) = 3x + 6$. Find: $(f - g)(3)$.

Solution: Use this: $(f - g)(x) = f(x) - g(x)$.

Then: $(f - g)(x) = (5x^2 - 3) - (3x + 6) = 5x^2 - 3 - 3x - 6 = 5x^2 - 3x - 9$.

Substitute x with 3: $(f - g)(3) = 5(3)^2 - 3(3) - 9 = 45 - 9 - 9 = 27$.

Example 5. If $f(x) = 1 + 2x$ and $g(x) = -x + 1$. Find $(g - f)(-3a)$.

Solution: $(g - f)(x) = (-x + 1) - (1 + 2x) = -x + 1 - 1 - 2x = -3x$.

Substitute x with $-3a$: $(g - f)(-3a) = -3(-3a) = 9a$.

Multiplying and Dividing Functions

- Just like we can multiply and divide numbers and expressions, we can multiply and divide two functions and simplify or evaluate them.

- For two functions $f(x)$ and $g(x)$, we can create two new functions:

$$(f \cdot g)(x) = f(x) \cdot g(x) \text{ and } \left(\frac{f}{g}\right)(x) = \frac{f(x)}{g(x)}$$

Examples:

Example 1. If $g(x) = x + 3$, $f(x) = x + 4$. Find: $(g \cdot f)(x)$.

Solution: According to: $(f \cdot g)(x) = f(x) \cdot g(x)$.

So: $(x + 3)(x + 4) = x^2 + 4x + 3x + 12 = x^2 + 7x + 12$.

Example 2. If $f(x) = x + 6$, and $h(x) = x - 9$. Find: $\left(\frac{f}{h}\right)(x)$.

Solution: Use this: $\left(\frac{f}{g}\right)(x) = \frac{f(x)}{g(x)}$. We have: $\left(\frac{f}{h}\right)(x) = \frac{x+6}{x-9}$.

Example 3. If $g(x) = x + 7$, and $f(x) = x - 3$. Find: $(g \cdot f)(2)$.

Solution: $(g \cdot f)(x) = (x + 7)(x - 3) = x^2 - 3x + 7x - 21 = x^2 + 4x - 21$.

Substitute x with 2: $(g \cdot f)(2) = (2)^2 + 4(2) - 21 = 4 + 8 - 21 = -9$.

Example 4. If $f(x) = x + 3$, and $h(x) = 2x - 4$. Find: $\left(\frac{f}{h}\right)(3)$.

Solution: We have: $\left(\frac{f}{h}\right)(x) = \frac{f(x)}{h(x)} = \frac{x+3}{2x-4}$.

Substitute x with 3: $\left(\frac{f}{h}\right)(3) = \frac{3+3}{2(3)-4} = \frac{6}{2} = 3$.

Example 5. If $f(x) = 2x + 1$ and $g(x) = -2x + 1$, find the value of x for which: $(f \cdot g)(x) = (f + g)(x)$.

Solution: $(2x + 1)(-2x + 1) = (2x + 1) + (-2x + 1)$.

So: $-4x^2 + 2x - 2x + 1 = 2$, which we can simply further:

$-4x^2 + 1 = 2 \Rightarrow -4x^2 = 1 \Rightarrow x^2 = -\frac{1}{4}$.

Since we can't have a real number squared that equals a negative value, there's no real value of x for which: $(f \cdot g)(x) = (f + g)(x)$.

Composition of Functions

- "Composition of functions" simply means combining two or more functions in a way where the output from one function becomes the input for the other function.

  The notation used for composition is: $(fog)(x) = f(g(x))$ and is read "f composed with g of x" or "f of g of x".

Examples:

Example 1. Using $f(x) = 2x + 3$ and $g(x) = 5x$, find: $(fog)(x)$.

Solution: Using definition: $(fog)(x) = f(g(x)) = f(5x)$.

Now, find $f(5x)$ by substituting x with $5x$ in $f(x)$ function.

$f(x) = 2x + 3 \Rightarrow f(5x) = 2(5x) + 3 = 10x + 3$.

Example 2. Using $f(x) = 3x - 1$ and $g(x) = 2x - 2$, find: $(gof)(5)$.

Solution: $(gof)(x) = g(f(x)) = g(3x - 1)$.

Now, substitute x in $g(x)$ by $(3x - 1)$:

$g(3x - 1) = 2(3x - 1) - 2 = 6x - 2 - 2 = 6x - 4$.

Substitute x with 5: $(gof)(5) = 6(5) - 4 = 30 - 4 = 26$.

Example 3. Using $f(x) = 2x^2 - 5$ and $g(x) = x + 3$, find: $f(g(3))$.

Solution: First, find $g(3)$: $g(x) = x + 3 \Rightarrow g(3) = 3 + 3 = 6$.

$f(g(3)) = f(6) = 2(6)^2 - 5 = 2(36) - 5 = 67$.

Example 4. Using $f(x) = 3x - 5$ and $g(x) = -2x$, find $g\big(f(g(x))\big)$.

Solution: First, we find the inner composite, $f(g(x))$:

$f(g(x)) = 3(-2x) - 5 = -6x - 5$, and then, $g\big(f(g(x))\big)$:

$$g\big(f(g(x))\big) = g(-6x - 5) = -2(-6x - 5) = 12x + 10$$

Writing Functions

- A function is a kind of relationship between variables, and it shows how 2 things are related to each other. It can take various forms, like an equation.

- Every function consists of inputs and outputs. The input is a variable that goes into the function it is also called the independent variable or **domain**. The output is a variable that comes out of the function, and it is also called the dependent variable or **range**.

- The function rule is an algebraic statement that specifies a function. The function determines which inputs are suitable, and the outputs will be determined by the inputs. 3 methods to show functions are graphs, tables, and algebraic expressions.

- A table that shows a function usually includes a column of inputs, a column of outputs, and a third column between the input and output to present how the outputs can be made by the inputs.

Examples:

Example 1. According to the values of x and y in the following relationship, find the right equation. $\{(1,4),(2,8),(3,12),(4,16)\}$

Solution: Find the relationship between the first x-value and first y-value: $(1,4)$. The value of y is 4 times the value of x: $1 \times 4 = 4$. The equation is $y = 4x$. Now draw the input-output table of the values to make sure the relationship is correct all the way:

x	$y = 4x$	y
1	$4(1) = 4$	4
2	$4(2) = 8$	8
3	$4(3) = 12$	12
4	$4(4) = 16$	16

Example 2. According to the values of x and y in the following relationship, find the right equation: $\{(1,3),(2,4),(3,5),(4,6)\}$

Solution: Find the relationship between the first x-value and first y-value: $(1,3)$. The value of y is 2 more than the x-value: $1 + 2 = 3$. So, the equation is $y = x + 2$. Now check the other values: $2 + 2 = 4, 3 + 2 = 5, 4 + 2 = 6$. Therefore each x-value and y-value satisfies the equation $y = x + 2$.

EffortlessMath.com

Graphing Functions

- Graphing functions is a way to show the function on the coordinate plane and to achieve this purpose, you can draw a line or a curve that represents a given function on the coordinate plane. A graph represents a function when each point on the line (or curve) of the graph can be an answer to the function equation.

- To graph linear functions and quadratic functions, first determine the shape of the graph. If it's a linear function and it's in form $f(x) = ax + b$, then this graph should be a line. If it's a quadratic function and it's in form $f(x) = ax^2 + bx + c$, then this graph should be a parabola.

- To graph linear functions, consider some random points on it, then substitute some random x-values in the equation and find the related y-values of the equation, to be able to graph the ordered pairs.

- To graph a quadratic function, you can consider some random points on it. But the graph of a quadratic function is a perfect U-shape, so to find a perfect U-shaped parabola, you should find its vertex. This means finding the point that the curve is turning. The next step after finding the vertex is finding 2 or 3 random points on every side of the vertex and then graphing the parabola.

Example:

Graph the following function: $y = 3x - 2$.

Solution: First, consider some random x-values, make an input-output table, and find their ordered pairs. These ordered pairs satisfy the function. Now, use the ordered pairs and draw a line that passes through all the points.

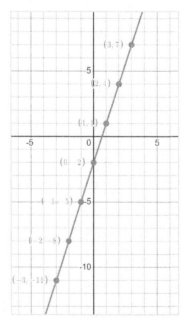

x	$y = 3x - 2$	(x, y)
-3	$3(-3) - 2 = -11$	$(-3, -11)$
-2	$3(-2) - 2 = -8$	$(-2, -8)$
-1	$3(-1) - 2 = -5$	$(-1, -5)$
0	$3(0) - 2 = -2$	$(0, -2)$
1	$3(1) - 2 = 1$	$(1, 1)$
2	$3(2) - 2 = 4$	$(2, 4)$
3	$3(3) - 2 = 7$	$(3, 7)$

bit.ly/3wt8sNP

Parent Functions

- The parent function is the simplest form of representation of any function without transformations.

- The most important parent functions include constant, linear, absolute-value, polynomials, rational, radical, exponential, and logarithmic functions.

- For the function $f(x)$ and constant number $k > 0$, the transformations of functions in which the properties of the parent function are preserved is as follows:

$y = f(x) + k$	$y = f(x - k)$	$y = -f(x)$
If $k > 0$, shifted up If $k < 0$, shifted down	If $k > 0$, shifted to the right If $k < 0$, shifted to the left	is symmetric to the function $y = f(x)$ with respect to the x−axis.
$y = kf(x)$	$y = f(kx)$	$y = f(-x)$
If $k > 1$, $f(x)$ is stretched vertically. If $0 < k < 1$, $f(x)$ is compressed vertically	If $k > 1$, $f(x)$ is compressed horizontally. If $0 < k < 1$, $f(x)$ is stretched horizontally	is symmetric to the function $y = f(x)$ with respect to the y−axis.

Example:

What is the parent graph of the following function and what transformations have taken place on it:

$$y = 2x^2 - 1$$

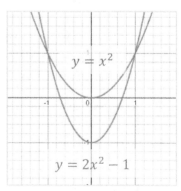

Solution: We know that the graph $y = f(x) + k$; $k < 0$, is shifted k units to the down of the graph $y = f(x)$. Then, $y = 2x^2 - 1$, is shifted 1 unit down of the graph $y = 2x^2$.

On the other hand, $y = kf(x)$; $k > 1$ is stretched vertically then the function $y = x^2$ is stretched vertically by a factor of 2.

Clearly, you can see that the function $y = x^2$ is the parent function of $y = 2x^2 - 1$.

EffortlessMath.com

Function Inverses

- An inverse function is a function that reverses another function: If the function f applied to an input x gives a result of y, then applying its inverse function g to y gives the result x. $f(x) = y$ if and only if $g(y) = x$.
- The inverse function of $f(x)$ is usually shown by $f^{-1}(x)$.
- $f(x)$ is symmetric to $f^{-1}(x)$ with respect to the line $y = x$
- $f^{-1}(f(x)) = x$, meaning inverse function applied to the function, returns original input.
- To find the inverse of a function, we need to switch the placement of x and y, and then solve for the new y, which is actually the $f^{-1}(x)$.

Examples:

Example 1. Find the inverse of the function: $f(x) = 2x - 1$.

Solution: First, replace $f(x)$ with y: $y = 2x - 1$. Then, replace all x's with y and all y's with x: $x = 2y - 1$.

Now, solve for y: $x = 2y - 1 \Rightarrow x + 1 = 2y \Rightarrow \frac{1}{2}x + \frac{1}{2} = y$.

Finally replace y with $f^{-1}(x)$: $f^{-1}(x) = \frac{1}{2}x + \frac{1}{2}$.

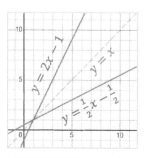

Example 2. Find the inverse of the function: $g(x) = \frac{1}{5}x + 3$.

Solution: Replace $g(x)$ with y: $g(x) = \frac{1}{5}x + 3 \Rightarrow y = \frac{1}{5}x + 3$.
Then, replace all x's with y and all y's with x: $x = \frac{1}{5}y + 3$.
Now, solve for y: $x - 3 = \frac{1}{5}y \Rightarrow 5(x - 3) = y \Rightarrow y = 5x - 15$.
Therefore, $g^{-1}(x) = 5x - 15$.

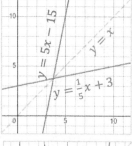

Example 3. Find the inverse of the function: $h(x) = \sqrt{x} + 6$.

Solution: $h(x) = \sqrt{x} + 6 \Rightarrow y = \sqrt{x} + 6$, replace all x's with y and all y's with x:

$x = \sqrt{y} + 6 \Rightarrow x - 6 = \sqrt{y} \Rightarrow (x - 6)^2 = (\sqrt{y})^2$

$\Rightarrow x^2 - 12x + 36 = y \Rightarrow h^{-1}(x) = x^2 - 12x + 36$.

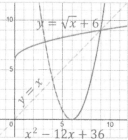

(Notice how the curve of $h(x)$ is only half as $h^{-1}(x)$ is. That's because a function has to have only one y for each x.)

Inverse Variation

- Inverse variation is a type of proportion in which one quantity decreases while another increases or vice versa. This means that the amount or absolute value of one quantity decreases if the other quantity increases so that their product always remains the same. This product is also known as the proportionality constant.

- An inverse variation is represented by the equation $xy = k$ or $y = \frac{k}{x}$, where $x \neq 0$, $y \neq 0$ and $k \neq 0$. $k = 0$ would nullify the inverse relationship.

- If **inverse variation** includes the points (x_1, y_1) and (x_2, y_2), it can be represented by the equation $x_1 \cdot y_1 = k$ and $x_2 \cdot y_2 = k$. (x_1, y_1) represents a specific point on the graph of an inverse variation. If you know one point on the graph, you can find the constant k by multiplying its coordinates: $x_1 \cdot y_1 = k$. Once you have the k, you can predict other points on the graph.

Examples:

Example 1. If y varies inversely as x and $y = 12$ when $x = 5$. What's the value of y when x is 3?

Solution: Use $y = 12$ and $x = 5$ to find the value of k:

$xy = k \Rightarrow k = 5 \times 12 \Rightarrow k = 60$.

Now that we have found k, we can again use the value of k, which is 60, and the value of x, which is 3, to get the value of y: $k = xy \Rightarrow 60 = 3 \times y \Rightarrow y = 20$.

Example 2. An inverse variation includes points $(4, 9)$ and $(12, m)$. Find m.

Solution: First find the k. Plug $x_1 = 4$ and $y_1 = 9$ in to the equation $x_1 y_1 = k$ and then solve for k: $x_1 y_1 = k \Rightarrow k = 4 \times 9 = 36$. Now use the inverse variation equation $x_2 y_2 = k$ to find m or y_2 when $x_2 = 12$: $x_2 y_2 = k \Rightarrow 12 y_2 = 36 \Rightarrow y_2 = 3$.

Example 3. If y varies inversely as x and $y = 0.5$ when $x = 8$. What's the value of x when y is 2?

Solution: $y_1 = 0.5$ and $x_1 = 8$, find k: $x_1 y_1 = k \Rightarrow k = 8 \times 0.5 = 4$.

$k = x_2 y_2 \Rightarrow 4 = x_2 \times 2 \Rightarrow x_2 = \frac{4}{2} \Rightarrow x_2 = 2$.

Domain and Range of functions

- For the function $f(x)$:
  - The set of all possible inputs is called the domain of the function.
  - The set of all possible outputs for each value of the domain is called the range of the function.
- The following methods can be used to obtain the domain of functions:
  - The domain of a polynomial function is all real numbers.
  - The domain of a square root function of the form $y = \sqrt{f(x)}$ is all values of x where, $f(x) \geq 0$.
  - The domain of an exponential function is all real numbers.
  - The domain of a logarithmic function of the form $y = \log_a f(x)$ is all values of x where, $f(x) > 0$.
  - Domain of a rational function of the form $y = \frac{f(x)}{g(x)}$, is all value of x where, $g(x) \neq 0$.
  - The domain of a piecewise function is the union of all the smaller domains.
- The domain of a function on the graph is the shadow of the graph on the x–axis.
- The range of a function on the graph is the shadow of the graph on the y–axis.

Examples:

Example 1. Find the domain and range of the function $f(x) = \frac{1}{x-1}$.

Solution: Since the domain of the rational function is a set of all real numbers except for where the denominator is zero: $x - 1 = 0$, so: $x = 1$. Then, the domain is all real numbers except for 1, also shown as: $\mathbb{R} - \{1\}$.

To find the range of the $f(x)$, we know that no matter how large or small x becomes, $f(x)$ will never be equal to 0. So, the range of $f(x)$ is all real numbers except zero.

Example 2. Find the domain and range of the function $g(x) = \sqrt{1-x} + 2$.

Solution: For domain, find non-negative values for radicals: $1 - x \geq 0$. Then the domain of the function is $1 - x \geq 0 \Rightarrow x \leq 1$. For range, we know that $\sqrt{1-x} \geq 0$, so $\sqrt{1-x} + 2 \geq 0 + 2 \Rightarrow \sqrt{1-x} + 2 \geq 2$. Therefore: $g(x) \geq 2$.

Piecewise Function

- A piecewise function uses more than one formula to define domain values.
- In general, a piecewise function can be shown as follows:

$$f(x) = \begin{cases} f_1(x), & x \in \text{Domain of } f_1(x) \\ f_2(x), & x \in \text{Domain of } f_2(x) \\ \vdots \\ f_n(x), & x \in \text{Domain of } f_n(x) \end{cases}$$

Where $n \geq 2$, and the domain of the function is the union of all of the smaller domains.

- Absolute value functions can be shown as piecewise functions.

Examples:

Example 1. Graph: $f(x) = \begin{cases} 2, & x < 0 \\ x + 1, & x \geq 1 \end{cases}$.

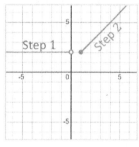

Solution: We know that solid point means include where the point is, and open or hollow point, means exclude. So:

Step 1: Graph $y = 2$ for $x < 0$. As follow:

Step 2: Graph $y = x + 1$ for $x \geq 1$.

Example 2. Write the piecewise function represented by the graph.

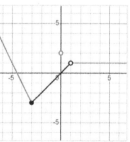

Solution: According to the graph, this function contains three pieces. The domain of the function is the union of three intervals as $(-\infty, -3]$, $[-3, 1)$, and $(1, +\infty)$. Using two points on the graph of each piece and point-slope form of the equation of a line, we write the linear function from them.

Step 1: The interval $(-\infty, -3]$, two points $(-4, -1)$ and $(-3, -3)$ lie on the line $y = -2x - 9$, because: $(y - y_0) = m(x - x_0) \Rightarrow (-3 + 1) = m(-3 + 4) \Rightarrow m = -2$. Then: $y = mx + b \Rightarrow -1 = -2(-4) + b \Rightarrow b = -9$, so $y = -2x - 9$.

Step 2: The interval $[-3, 1)$, two points $(-3, -3)$ and $(1, 1)$ lie on the line $y = x$.

Same as before: $1 + 3 = m(1 + 3) \Rightarrow m = 1 \Rightarrow b = 0$, so: $y = x$.

Step 3: The interval $(1, +\infty)$, the graph is a horizontal function passing through the point $(1, 1)$. Therefore: $f(x) = \begin{cases} -2x - 9 & x < -3 \\ x & -3 \leq x < 1 \\ 1 & x > 1 \end{cases}$.

Positive, Negative, Increasing and Decreasing Functions on Intervals

- For the function $f(x)$ over an arbitrary interval I:
  - f is called positive if for every x in the interval I, $f(x) > 0$. (Above the x−axis)
  - f is called negative if for every x in the interval I, $f(x) < 0$. (Under the x−axis)
- For the given function f, consider the points x and y in an arbitrary open interval of the domain. in this case:
  - The function f is increasing if for every $x < y$ in the interval: $f(x) \leq f(y)$.
  - The function f is decreasing if for every $x < y$ in the interval: $f(x) \geq f(y)$.
  - The function f is constant if for every $x < y$ in the interval: $f(x) = f(y) = C$, such that C is a constant number.

Example:

According to the graph, determine in which intervals the function is positive, negative, increasing, or decreasing.

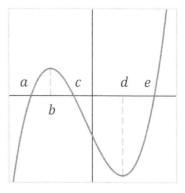

Solution:

Considering the graph, you see that:

- For $(-\infty, a)$ and (c, e), the graph is under the x−axis.
  So, the function corresponding to the graph is negative.
- For (a, c) and $(e, +\infty)$, the graph is above the x−axis.
  So, the function is positive.
- Since the values of the function increase in the intervals $(-\infty, b)$ and $(d, +\infty)$, the corresponding function of the graph is increasing in each interval.

Chapter 1: Practices

✎ Evaluate each function.

1) $g(n) = 2n + 5$, find $g(2)$

2) $h(n) = 5n - 9$, find $h(4)$

3) $k(n) = 10 - 6n$, find $k(2)$

4) $g(n) = -5n + 6$, find $g(-2)$

5) $k(n) = -8n + 3$, find $k(-6)$

6) $w(n) = -2n - 9$, find $w(-5)$

✎ Perform the indicated operation.

7) $f(x) = x + 6$
   $g(x) = 3x + 2$
   Find $(f - g)(x)$

8) $g(x) = x - 9$
   $f(x) = 2x - 1$
   Find $(g - f)(x)$

9) $h(x) = 5x + 6$
   $g(x) = 2x + 4$
   Find $(h + g)(x)$

10) $g(x) = -6x + 1$
    $f(x) = 3x^2 - 3$
    Find $(g + f)(5)$

11) $g(x) = 7x - 1$
    $h(x) = -4x^2 + 2$
    Find $(g - h)(-3)$

12) $h(x) = -x^2 - 1$
    $g(x) = -7x - 1$
    Find $(h - g)(-5)$

Perform the indicated operation.

13) $g(x) = x + 3$
    $f(x) = x + 1$
    Find $(g \cdot f)(x)$
    _____

14) $f(x) = 4x$
    $h(x) = x - 6$
    Find $(f \cdot h)(x)$
    _____

15) $g(a) = a - 8$
    $h(a) = 4a - 2$
    Find $(g \cdot h)(3)$
    _____

16) $f(x) = 6x + 2$
    $h(x) = 5x - 1$
    Find $\left(\frac{f}{h}\right)(-2)$
    _____

17) $f(x) = 7x - 1$
    $g(x) = -5 - 2x$
    Find $\left(\frac{f}{g}\right)(-4)$
    _____

18) $g(a) = a^2 - 4$
    $f(a) = a + 6$
    Find $\left(\frac{g}{f}\right)(-3)$
    _____

Using $f(x) = 4x + 3$ and $g(x) = x - 7$, find:

19) $g(f(2)) =$ _____
20) $g(f(-2)) =$ _____
21) $f(g(4)) =$ _____
22) $f(f(7)) =$ _____
23) $g(f(5)) =$ _____
24) $g(f(-5)) =$ _____

According to the values of x and y in the following relationship, find the right equation.

25) $\{(1,3), (2,6), (3,9), (4,12)\}$ _____
26) $\{(1,5), (2,7), (3,9), (4,11)\}$ _____

Effortless Math Education

Chapter 1: Functions

📖 **What is the parent graph of the following function and what transformations have taken place on it?**

27) $y = x^2 - 3$

28) $y = x^3 + 4$

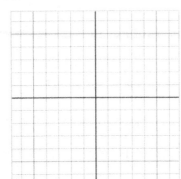
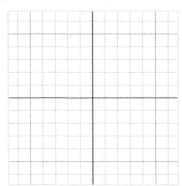

📖 **Find the inverse of each function.**

29) $f(x) = \frac{1}{x} - 6 \Rightarrow f^{-1}(x) =$

30) $g(x) = \frac{7}{-x-3} \Rightarrow g^{-1}(x) =$

31) $h(x) = \frac{x+9}{3} \Rightarrow h^{-1}(x) =$

32) $h(x) = \frac{2x-10}{4} \Rightarrow h^{-1}(x) =$

33) $f(x) = \frac{-15+x}{3} \Rightarrow f^{-1}(x) =$

34) $s(x) = \sqrt{x} - 2 \Rightarrow s^{-1}(x) =$

📖 **Solve.**

35) If y varies inversely as x and $y = 18$ when $x = 6$. What's the value of y when x is 4? _____

36) If y varies inversely as x and $y = 0.8$ when $x = 6$. What's the value of x when y is 3? _____

Effortless Math Education

✎ **Graph the following functions.**

37) $y = 2x - 5$

38) $y = 4x + 3$

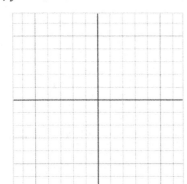
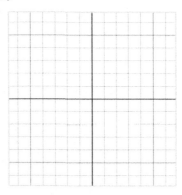

✎ **Find the domain and range of the functions.**

39) $y = x^3 - 4$

Domain: ____, Range: ____

40) $y = \sqrt{x - 8} + 4$

Domain: ____, Range: ____

41) $y = \frac{2}{2x-1}$

Domain: ____, Range: ____

42) $y = -2x^3 + 6$

Domain: ____, Range: ____

✎ **Graph.**

43) $f(x) = \begin{cases} 4 - 2, & x < -1 \\ x - 1, & x \geq 0 \end{cases}$

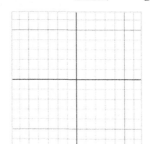

✎ **Solve.**

44) Determine the intervals where the function is increasing and decreasing. Submit your solution in interval notation.

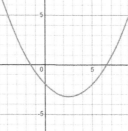

Effortless Math Education

Chapter 1: Answers

1) 9
2) 11
3) -2
4) 16
5) 51
6) 1
7) $-2x + 4$
8) $-x - 8$
9) $7x + 10$
10) 43
11) 12
12) -60
13) $x^2 + 4x + 3$
14) $4x^2 - 24x$
15) -50
16) $\frac{10}{11}$
17) $-\frac{29}{3}$
18) $\frac{5}{3}$
19) 4
20) -12
21) -9
22) 127
23) 16
24) -24
25) $y = 3x$
26) $y = 2x + 3$

27) Parent: Quadratic

Transformations: Down 3

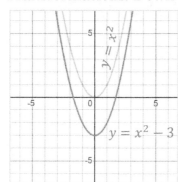

28) Parent: Cubic

Transformations: Up 4

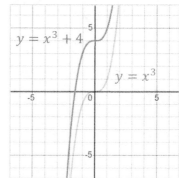

29) $f^{-1}(x) = \frac{1}{x+6}$

30) $g^{-1}(x) = -\frac{7+3x}{x}$

31) $h^{-1}(x) = 3x - 9$

32) $h^{-1}(x) = 2x + 5$

33) $f^{-1}(x) = 3x + 15$

34) $s^{-1}(x) = x^2 + 4x + 4$

35) $y = 27$

36) $x = 1.6$

37)
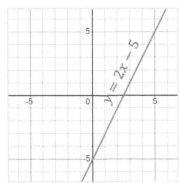

38)
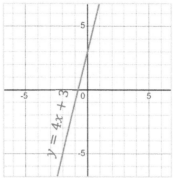

39) Domain: $-\infty < x < \infty$

Range: $-\infty < f(x) < \infty$

40) Domain: $x \geq 8$

Range: $f(x) \geq 4$

41) Domain: $x < \frac{1}{2}$ or $x > \frac{1}{2}$

Range: $f(x) < 0$ or $f(x) > 0$

42) Domain: $-\infty < x < \infty$

Range: $-\infty < f(x) < \infty$

43)
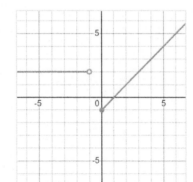

44) Increasing intervals: $(3, +\infty)$

Decreasing intervals: $(-\infty, 3)$

CHAPTER

2 Advanced Functions

Math topics that you'll learn in this chapter:

- ☑ Exponential Functions
- ☑ Linear, Quadratic and Exponential Models
- ☑ Linear vs Exponential Growth
- ☑ Logarithms and Natural Logarithms
- ☑ Sin, Cos, Tan, Cot, Sec, Csc
- ☑ Domain and Range of Trigonometric Functions
- ☑ Trigonometric Functions Values for key angles
- ☑ The Unit Circle and Trigonometric Identities
- ☑ Floor and Ceiling Functions

Exponential Function

- An exponential function with a base a is shown as $f(x) = a^x$, where a is a positive real number, and $a \neq 1$.

- Exponential Growth and Decay:
    - If $a > 1$, the function $f(x) = a^x$ is exponential growth.
    - If $0 < a < 1$, the function $f(x) = a^x$ is exponential decay.

- All exponential functions of the form $f(x) = a^x$ have the same domain, range and y-intercept. As follow: Domain: $\mathbb{R}$, Range: $(0, +\infty)$, y-intercept: 1.

Example:

Graph $f(x) = 2^x$, and $g(x) = -5f(x) + 1$. Which one is exponential growth or decay? Then give domain, range and y-intercept.

Solution: Since the base of the $f(x) = 2^x$ is 2 and greater than 1, then $f(x)$ is growth with domain $\mathbb{R}$, range $(0, +\infty)$, and y-intercept 1.

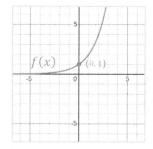

All values of $y = 5f(x)$ are 5 times the function $y = f(x)$, and the function $y = 5f(x)$ is symmetric to the function $y = -5f(x)$ with respect to the x-axis. As follow:

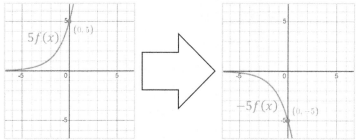

On the other hand, $y = -5f(x) + 1$ is shifted 1 unit to the up, therefore, according to the graph of $y = -5f(x) + 1$, the y-intercept is -4, the domain is $\mathbb{R}$, and the range is $(-\infty, 1)$. And the function is exponential decay.

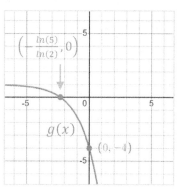

Linear, Quadratic and Exponential Models

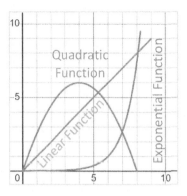

- Functions are a kind of mathematical relationship between inputs and outputs and linear, quadratic, and exponential functions are the 3 main types of functions.

- A linear function is a type of function whose highest exponent is 1 and the standard form of this function is $y = mx + b$. The graph of the linear function shows a straight line; Therefore, its name is a linear function.

- Quadratic functions when graphed make parabolas. A quadratic function is a type of polynomial function whose highest exponent is 2. The quadratic function's standard form is $y = ax^2 + bx + c$.

- The exponential function is a type of function that includes variables in the exponent. An exponential function's general form is $y = e^x$.

- The graphs of linear functions are straight lines with no curve. The graphs of quadratic functions are parabola shaped. The graphs of exponential function have a curve. This curve can be vertical in the beginning and then grow to be horizontal or can be horizontal in the beginning and then become more vertical.

- You'll be able to find the degree of the model for a given ordered pair's information by determining the differences between dependent values:
  - The model is linear if the first difference is constant.
  - The model is quadratic if the second difference is constant.
  - The model is exponential if the independent variable changes by a constant ratio.

Example:

Determine whether the following table of values represents a linear function, an exponential function, or a quadratic function.

x	y
-5	-5
-3	-2
-1	1
1	4
3	7

Solution: The model is related to a linear function because the first difference is the same value:

$y \Rightarrow -2 - (-5) = 3, 1 - (-2) = 3, 4 - 1 = 3, 7 - 4 = 3.$

$x \Rightarrow -3 - (-5) = 2, -1 - (-3) = 2, 1 - (-1) = 2, 3 - 1 = 2.$

Linear vs Exponential Growth

- The difference between linear and exponential growth is related to the difference in the y values change when the x values increase by a constant amount. In such a way that:
    - A relationship is a linear growth if the y values have equal differences.
    - A relationship is an exponential growth if the y values have an equal ratio.
    - Otherwise, it is neither.
- Actually, a linear growth diagram is similar to a straight line and an exponential growth diagram is similar to an exponential function.
- Much faster than linear growth is exponential growth.

Examples:

Example 1. If 75 students graduate from high school every year. Is the relationship linear, exponential, or neither?

List the data as follows: (1,75), (2,150), (3,225), (4,300) and ⋯.

Solution: Considering that in each year first components increase by exactly 1 unit and the value of the second component increases with a constant difference of 75. It is Linear growth.

Example 2. Using the data in this table, determine whether this relationship is linear, exponential, or neither.

x	0	1	2	3	4	5
y	3	6	12	24	48	96

Solution: We can see that the x value increase by 1 unit and the y value increase by the constant ratio of 2. Then, this relationship is exponential.

Example 3. According to the following diagram, determine whether this relationship is linear, exponential, or neither.

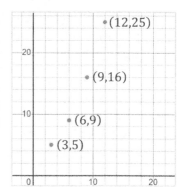

Solution: Here, the x value increase by 3 unit, however, the difference between the y values are not constant, and the ratios aren't constant. Therefore, it is neither.

Logarithms

- In mathematics, logarithms streamline calculations, illuminate exponential concepts, and find applications in financial computations, astronomy for measuring stellar magnitudes, and in chemistry, to gauge acidity or alkalinity through pH values.

- Logarithm is another way of writing exponent. $\log_b y = x$ is equivalent to $y = b^x$.

$$\log_b y = x \Longrightarrow b^x = y$$

- The domain of logarithmic functions are positive values, and the range is $\mathbb{R}$ (or all real numbers).

- If the logarithm had no base, it's called the common logarithm, which always has a base of 10.

- Some basic logarithm Properties: ($a > 0$, $a \neq 0$, $M > 0$, $N > 0$, $k \in \mathbb{R}$)

log 0	Undefined
log 1	0
log 2	0.301
log 3	0.477
log 4	0.602
log 5	0.698
log 6	0.778
log 7	0.845
log 8	0.903
log 9	0.954
log 10	1

- $\log_a 1 = 0$
- $a^{\log_a k} = k$
- $\log_a a = 1$
- $\log_a x = \dfrac{1}{\log_x a}$
- $\log_b[f(x)] = \log_b[g(x)] \Rightarrow f(x) = g(x)$ **(And vice versa)**
- $\log_a(M \times N) = \log_a M + \log_a N$ **(Product Rule)**
- $\log_a \dfrac{M}{N} = \log_a M - \log_a N$ **(Quotient Rule)**
- $\log_a M^k = k \log_a M$ **(Power Rule)**
- $\log_x y = \dfrac{\log_a y}{\log_a x}$ **(Change of base)**

Examples:

Example 1. Evaluate: $\log_2 32$.
Solution: We can solve this logarithm by power rule: $\log_2 32 = \log_2 2^5$.
So: $\log_2 2^5 = 5 \log_2 2 = 5 \times 1 = 5$.
(We can also assign the arbitrary variable k as the answer to this logarithm:
So: $\log_2 32 = k$, and from definition, we know that: $2^k = 32$. Solving for k: $k = 5$)

Example 2. Evaluate: $\log_x \sqrt{x^3}$.
Solution: We can rewrite $\sqrt{x^3}$ as $x^{3/2}$, so:
$\log_x \sqrt{x^3} = \log_x x^{\frac{3}{2}} = \frac{3}{2} \log_x x = \frac{3}{2} \times 1 = \frac{3}{2}$.

Properties of logarithms

- Using some of the properties of logs mentioned earlier, sometimes we can expand a logarithm expression (expanding) or convert some logarithm expressions into a single logarithm (condensing).
- Here are some additional logarithm rules you can use:
  - $\log_a x = \log_{a^c} x^c$
  - $\log_{a^k} x = \frac{1}{k} \log_a x, k \neq 0$

Examples:

Example 1. Expand this logarithm. $\log_a(15)$

Solution: Use log rule: $\log_a(x \times y) = \log_a x + \log_a y$.

Then: $\log_a(3 \times 5) = \log_a 3 + \log_a 5$.

Example 2. Condense this expression to a single logarithm. $\log_a 2 - \log_a 7$

Solution: Use log rule: $\log_a x - \log_a y = \log_a \frac{x}{y}$. Then: $\log_a 2 - \log_a 7 = \log_a \frac{2}{7}$.

Example 3. Simplify: $2 \log_{10} \sqrt{2} + \log_{10} 5$.

Solution: We can rewrite the radical as $2^{\frac{1}{2}}$ and go from there, but we can also send the coefficient (2) as an exponent for the radical, so we have:

$\log_{10} \sqrt{2}^2 + \log_{10} 5 = \log_{10} 2 + \log_{10} 5$, which can be merged using product rule:

$$\log_{10} 2 + \log_{10} 5 = \log_{10} 10 = 1$$

Example 4. Evaluate the composite logarithm: $\log_3(\log_3(\log_2 8))$.

Solution: We start from the core: $\log_2 8 = \log_2 2^3 = 3 \log_2 2 = 3$, moving on to the next logarithm: $\log_3(\log_2 8) = \log_3 3 = 1$, lastly: $\log_3(\log_3(\log_2 8)) = \log_3 1$, which from the rule: $\log_a 1 = 0$, the answer is going to be 0.

Example 5. Evaluate $(\sqrt{10})^{\frac{2}{3} \log 5}$.

Solution: $(\sqrt{10})^{\frac{2}{3} \log 5} = (10^{\frac{1}{2}})^{\frac{2}{3} \log 5} = 10^{\frac{1}{3} \log 5} = 10^{\log 5^{\frac{1}{3}}} = 10^{\log \sqrt[3]{5}} = 10^{\log_{10} \sqrt[3]{5}}$, and from $a^{\log_a k} = k$: $10^{\log_{10} \sqrt[3]{5}} = \sqrt[3]{5}$.

Natural Logarithms

- A natural logarithm is a logarithm that has a special base of the mathematical constant e (which is an irrational number approximately equal to 2.71).

- The natural logarithm of x is generally written as $\ln x$, or $\log_e x$. So: $\ln x = \log_e x$.

- e is also called "the Neper number", "Euler's number" or "exponential constant", because it's central in exponential growth/decay equations.

$\ln 0$	Undefined
$\ln 1$	0
$\ln 2$	0.69
$\ln 5$	1.6
$\ln 10$	2.3
$\ln 100$	4.6
$\ln 500$	6.21
$\ln 1000$	6.9
$\ln \infty$	∞

Examples:

Example 1. Expand this natural logarithm. $\ln 4x^2$
Solution: Use log rule: $\log_a(x \times y) = \log_a x + \log_a y$.
Therefore: $\ln 4x^2 = \ln 4 + \ln x^2$.
Now, use log rule: $\log_a(M)^k = k \log_a(M)$. Then: $\ln 4 + \ln x^2 = \ln 4 + 2 \ln x$.

Example 2. Condense this expression to a single logarithm. $\ln x - \log_e 2y$
Solution: Use log rule: $\log_a x - \log_a y = \log_a \frac{x}{y}$. Then: $\ln x - \log_e 2y = \ln \frac{x}{2y}$.

Example 3. Solve this equation for x: $e^x = 6$.
Solution: If $f(x) = g(x)$, then: $\ln(f(x)) = \ln(g(x)) \Rightarrow \ln(e^x) = \ln(6)$.
Use log rule: $\log_a(M)^k = k \log_a(M) \Rightarrow \ln(e^x) = x \ln(e) \Rightarrow x \ln(e) = \ln(6)$.
We know that: $\ln(e) = \log_e e = 1$, then: $x = \ln(6)$.

Example 4. Solve this equation for x: $\ln(4x - 2) = 1$.
Solution: Use log rule: $a = \log_b(b^a) \Rightarrow 1 = \ln(e^1) = \ln(e) \Rightarrow \ln(4x - 2) = \ln(e)$.
When the logs have the same base: $\log_b(f(x)) = \log_b(g(x)) \Rightarrow f(x) = g(x)$.
$\ln(4x - 2) = \ln(e)$, then: $4x - 2 = e \Rightarrow 4x = e + 2 \Rightarrow x = \frac{e+2}{4}$.

Example 5. Solve this equation for x: $\ln(3x + 5) = 2$.
Solution: Raise e (the base of natural logarithms) to the power of both sides:
$e^{\ln(3x+5)} = e^2$, and we know that $a^{\log_a k} = k$, so: $3x + 5 = e^2$.

$$3x = e^2 + 5 \Rightarrow x = \frac{e^2 - 5}{3}$$

Sine, Cosine, and Tangent

- Sine, cosine, and tangent are fundamental trigonometric functions used in mathematics to relate the angles of a right triangle to the lengths of its sides. These functions are essential for understanding various mathematical concepts, including geometry, calculus, and physics, and analyzing waves, vibrations, and oscillations.

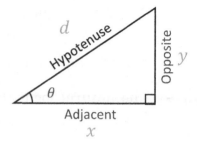

- **Sine (*sin*)**: The sine function relates the ratio of the length of the side opposite to an angle in a right triangle to the length of the hypotenuse. It is defined as $sin(\theta) = \frac{opposite}{hypotenuse} = \frac{y}{d}$, where θ is the angle.
- **Cosine (*cos*)**: The cosine function relates the ratio of the length of the side adjacent to an angle in a right triangle to the length of the hypotenuse. It is defined as $cos(\theta) = \frac{adjacent}{hypotenuse} = \frac{x}{d}$, where θ is the angle.
- **Tangent (*tan*)**: The tangent function is the ratio of the sine function to the cosine function (or the ratio of the length of the side opposite to an angle in a right triangle to the length of the side adjacent to the angle). It is defined as $tan(\theta) = \frac{sin(\theta)}{cos(\theta)}$ or $\frac{opposite}{adjacent} = \frac{y}{x}$, where θ is the angle.

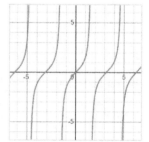

Example:

Find the value of Sine, Cosine, and Tangent of the angle $\widehat{BAC}$ in the figure below.

Solution: According to the size of the sides of the triangle and using the Sine, Cosine, and Tangent formulas, we will have:

$sin(\theta) = \frac{opposite}{hypotenuse} = \frac{6.81}{10.5} \cong 0.65$

$cos(\theta) = \frac{adjacent}{hypotenuse} = \frac{8}{10.5} \cong 0.76$

$tan(\theta) = \frac{opposite}{adjacent} = \frac{6.81}{8} \cong 0.85$

(or we could divide $sin(\theta)$ by $cos(\theta)$ and get the same result for $tan(\theta)$: $\frac{0.65}{0.76} \cong 0.85$)

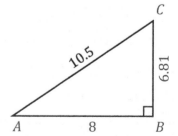

Reciprocal Functions: Cosecant, Secant, and Cotangent

- The trigonometric functions that can be defined in terms of $sin\,\theta$, $cos\,\theta$, and $tan\,\theta$ are called the reciprocal functions. They include sec, csc and cot.
- **Secant** is a trigonometric function defined for all real numbers of θ except where $cos(\theta) \neq 0$, with the formula: $sec(\theta) = \frac{1}{cos(\theta)}$. It is undefined at angles of θ where $\theta = \left(n + \frac{1}{2}\right)\pi$ for any integer n. (For example, at $n = 1$: $\theta = \frac{3}{2}\pi = \frac{3}{2} \times 180 = 270°$, and $cos(270°) = 0$)

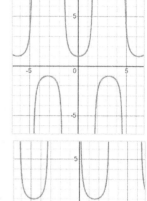

- Pi (π) is considered as 180 degrees because it represents half of a full rotation in a circle with a radius of 1. (Circumference: $2\pi r$)
- **Cosecant** is a trigonometric function defined for all real numbers of θ except where $sin(\theta) \neq 0$, with the formula: $csc(\theta) = \frac{1}{sin(\theta)}$. It is undefined at angles of θ where $\theta = n\pi$ for any integer n.

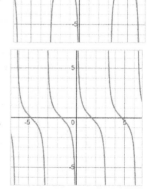

- **Cotangent** function is a trigonometric function defined for all real numbers of θ for which $tan(\theta)$ is not equal to 0, with the formula: $cot(\theta) = \frac{1}{tan(\theta)} = \frac{cos(\theta)}{sin(\theta)}$. It is undefined where $sin(\theta) = 0$, so it is defined for all θ except for $n\pi$, for any integer n.
- By dividing csc by sec: $\frac{csc(\theta)}{sec(\theta)} = \frac{\frac{1}{sin(\theta)}}{\frac{1}{cos(\theta)}} = \frac{cos(\theta)}{sin(\theta)} = cot(\theta)$.
- A function is even if $f(x) = f(-x)$, and it's odd if $f(x) = -f(-x)$. $sin(-x) = -sin(x)$, so it's an odd function, but $cos(-x) = cos(x)$, so it's even. Among trigonometric functions, only cos and sec are even.

Examples:

Example 1. Find the value of $sec(\theta)$ if $cos(\theta) = \frac{2}{7}$ using the reciprocal identity.

Solution: The reciprocal identity of sec is: $sec(\theta) = \frac{1}{cos(\theta)}$. If $cos(\theta) = \frac{2}{7}$, then $sec(\theta) = \frac{1}{\frac{2}{7}} = \frac{7}{2}$.

Example 2. Simplify the function $tan(\theta)\,cot(\theta)\,sin(\theta)$.

Solution: The reciprocal identity of $cot(\theta)$: $cot(\theta) = \frac{1}{tan(\theta)}$. $tan(\theta)\,cot(\theta)\,sin(\theta) = tan(\theta) \times \frac{1}{tan(\theta)} \times sin(\theta) = sin(\theta)$

Domain and Range of Trigonometric Functions

- The domain of the sine function and cosine function is the set of real numbers.
- The range of the sine function and cosine function is this set of real numbers: $[-1,1]$.
- The domain of the tangent function is the set of real numbers except for $\frac{\pi}{2} + n\pi$ for all integer values of n.
- The range of the tangent function is the set of all real numbers.
- The domain of the cotangent function is the set of real numbers except for $n\pi$ for all integer values of n.
- The range of the cotangent function is the set of all real numbers.
- The domain of the secant function is the set of real numbers except for $\frac{\pi}{2} + n\pi$ for all integer values of n.
- The range of the secant function is the set of real numbers $(-\infty, -1] \cup [1, +\infty)$.
- The domain of the cosecant function is the set of real numbers except for $n\pi$ for all integer values of n.
- The range of the cosecant function is the set of real numbers $(-\infty, -1] \cup [1, +\infty)$.

Examples:

Example 1. Find the range of $y = 4\tan(x)$.

Solution: The range of $y = 4\tan(x)$ is $(-\infty, +\infty)$.

Example 2. Find the domain and range of $y = \sin(x) - 4$.

Solution: The range of $\sin x$ is $[-1,1]$.

$-1 \leq \sin(x) \leq 1 \Rightarrow -1 - 4 \leq \sin(x) - 4 \leq 1 - 4 \Rightarrow -5 \leq y \leq -3$

The domain is $(-\infty, +\infty)$.

Example 3. Find the domain of $y = 3\cos(x) + 4$.

Solution: The domain of $y = 3\cos(x) + 4$ is $(-\infty, +\infty)$.

Trigonometric Function Values for Key Angles

- It is useful to remember the exact values of the trigonometric function summarized below:

θ	0°	30°	45°	60°	90°
$sin(\theta)$	0	$\frac{1}{2}$	$\frac{\sqrt{2}}{2}$	$\frac{\sqrt{3}}{2}$	1
$cos(\theta)$	1	$\frac{\sqrt{3}}{2}$	$\frac{\sqrt{2}}{2}$	$\frac{1}{2}$	0
$tan(\theta)$	0	$\frac{\sqrt{3}}{3}$	1	$\sqrt{3}$	undefined

Examples:

Example 1. Find the exact value of $sec(45°)$.

Solution: We know that $sec(\theta) = \frac{1}{cos(\theta)}$.

$sec(45°) = \frac{1}{cos(45°)} = \frac{1}{\frac{\sqrt{2}}{2}} = \frac{2}{\sqrt{2}} = \frac{2}{\sqrt{2}} \times \frac{\sqrt{2}}{\sqrt{2}} = \frac{2\sqrt{2}}{2} = \sqrt{2}$.

Example 2. Find the value of $(sin(30°) \cdot cos(60°))$.

Solution: $sin(30°) = \frac{1}{2}$ and $cos(60°) = \frac{1}{2}$. So, $sin(30°) \cdot cos(60°) = \frac{1}{2} \times \frac{1}{2} = \frac{1}{4}$.

Example 3. Find the value of $cos(30°) + tan(0°) + sin(60°)$.

Solution: $cos(30°) = \frac{\sqrt{3}}{2}$, $tan(0°) = 0$ and $sin(60°) = \frac{\sqrt{3}}{2}$.

$cos(30°) + tan(0°) + sin(60°) = \frac{\sqrt{3}}{2} + 0 + \frac{\sqrt{3}}{2} = 2\frac{\sqrt{3}}{2} = \sqrt{3}$.

Example 4. Find the value of $cos(sin(180°))$.

Solution: $sin(180°) = 0$, so $cos(sin(180°)) = cos(0°) = 1$.

The Unit Circle

- The unit circle is a fundamental concept in trigonometry. It's a circle with a radius of 1 centered at the origin in a Cartesian coordinate system.
- Key features include its equation, $x^2 + y^2 = 1$, and its use in defining trigonometric functions.

It aids in understanding angles, sine, cosine, and tangent.

- If $\angle POR$ is an angle in standard position and P is the point that the terminal side of the angle intersects the unit circle and $\angle POR = \theta$. Then:

  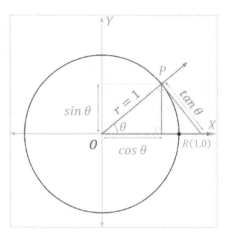

  - The sine function is a set of ordered pairs $(\theta, sin(\theta))$ that $sin(\theta)$ is the y coordinate of P.
  - The cosine function is the set of ordered pairs $(\theta, cos(\theta))$ that $cos(\theta)$ is the x-coordinate of P.

Examples:

Example 1. Determine the angle θ, in standard position, whose terminal side passes through this point on the unit circle: $P\left(\frac{\sqrt{3}}{2}, -\frac{1}{2}\right)$.

Solution: We have point P on the circle, in the 4th quadrant, with $x = \frac{\sqrt{3}}{2}$ and $y = -\frac{1}{2}$. So $-\frac{1}{2} = y = sin(\theta)$ and $\frac{\sqrt{3}}{2} = x = cos(\theta)$. We need to find an angle with a $sin(\theta) = \frac{1}{2}$ and $cos(\theta) = \frac{\sqrt{3}}{2}$. Using $arcsin$ or $arccos$: $\theta = -30°$.

Example 2. Does point $P\left(\frac{1}{4}, \frac{1}{4}\right)$ lie on the unit circle?

Solution: The equation of a unit circle is: $x^2 + y^2 = 1$. Now substitute $x = \frac{1}{4}$ and $y = \frac{1}{4}$: $\left(\frac{1}{4}\right)^2 + \left(\frac{1}{4}\right)^2 = \frac{1}{8} \neq 1$.

Since $x^2 + y^2 \neq 1$, the point $P\left(\frac{1}{4}, \frac{1}{4}\right)$ does not lie on the unit circle.

Additional Trigonometric Reminders

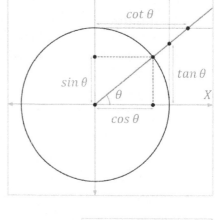

- Here is another representation of the relationship between unit circle and trigonometric functions.
- It can be helpful to visualize angles when we forget the trigonometric function value for an angle. For example, sine function is measured on y-axis. When θ is 90°, $sin(\theta) = 1$, or, at 0° and 180°, $sin(\theta) = 0$.
Similarly, $cos(0°) = 1$ and $cos(180°) = -1$ and so on. Meaning you only need to picture this circle and find the answers, or approximations.

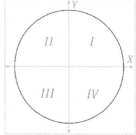

- The unit circle is comprised of 4 quarters, called "Quadrant". The first quadrant, located in the top right-hand position. Proceeding counterclockwise, as shown on this graph, we have the next quadrants. So, for the first quadrant, we have angles that are greater than or equal to 0° and less than 90°. And so on.
- For remembering the signs of each trigonometric value, you can use "all $-sin-tan/cot-cos$" or "astc" acronym, respectively for 1, 2, 3, 4 quadrants. (excluding sec and csc). For example, in 1st quadrant, all functions are positive, meaning for $0° < \theta < 90°$, sin, cos, tan, cot are positive. For 195° for example (3rd quadrant), tan and cot are positive and the other functions are not.
- Some very useful trigonometric identities can be assembled from the unit circle and Pythagorean theorem. For example: $sin^2 x + cos^2 x = 1$ is derived from the Pythagorean theorem applied to the unit circle. here are some other ones:
  - $1 + tan^2(x) = \frac{1}{cos^2(x)}$
  - $cot(2x) = \frac{cot^2(x-1)}{2\,cot(x)}$
  - $sin(2x) = 2\,sin(x)\,cos(x)$
  - $1 + cot^2(x) = \frac{1}{sin^2(x)}$
  - $tan(2x) = \frac{2\,tan(x)}{1-tan^2(x)}$
  - $cos(2x) = 1 - 2\,sin^2(x)$

Example:

If $sin(30°) = \frac{1}{2}$, find $cos(60°)$.

Solution: Using $cos(2\theta) = 1 - 2\,sin^2(\theta)$, we have: $cos(60°) = 1 - 2(sin(30°))^2$. After plugging the given number:
$cos(60°) = 1 - 2\left(\frac{1}{2}\right)^2 = 1 - 2\left(\frac{1}{4}\right) = 1 - \frac{1}{2} = \frac{1}{2}$.

Periodic Properties of Trigonometric Functions

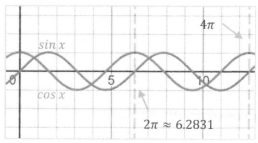

- The sine function is periodic with a period of 2π radians (or 360 degrees). Meaning the sine values repeat every 2π radians, or $2k\pi$ values ($k \in \mathbb{Z}$, meaning $\cdots, -2, -1, 0, 1, 2, \cdots$). For example, $sin(380°) = sin(360° + 20°) = sin(20°)$.
- The period of a function is the shortest distance it can be shifted without changing its shape or behavior.
- Cosine, tangent and cotangent have their own periods too. Here is the general formula for each function, for integer values of k:
  - $sin(x) = sin(2k\pi + x)$
  - $cos(x) = cos(2k\pi + x)$
  - $tan(x) = tan(k\pi + x)$
  - $cot(x) = cot(k\pi + x)$

  Similarly, for sine and cosine shifted by π instead of 2π, we have:
  - $sin(x) = -sin(\pi + x)$
  - $cos(x) = -cos(\pi + x)$
- When using periodic properties, it's essential to consider where a function is on the cycle. If it's below a period, we need to adjust $(2k\pi + x)$ in the formula accordingly, because $(2k\pi + x)$ means values greater than $2k\pi$, but the function is below $2k\pi$. For example, $sin(x) = sin(2k\pi + x)$, to find $sin(330°)$, we say $330° = 360° - 30°$ or $(360° + (-30°))$ or $(2k\pi + (-\frac{\pi}{6}))$, so in this case, instead of using formula as $sin(330°) = sin(30°)$, we need to write: $sin(330°) = sin(-30°) = -sin(30°)$, since sine is an odd function.
- These formulas allow you to express one trigonometric function in terms of another by utilizing the concept of complementary angles, and can be useful when converting angles of trigonometric functions:
  - $sin(x) = cos\left(\frac{\pi}{2} - x\right)$
  - $cos(x) = sin\left(\frac{\pi}{2} - x\right)$
  - $tan(x) = cot\left(\frac{\pi}{2} - x\right)$
  - $cot(x) = tan\left(\frac{\pi}{2} - x\right)$

Examples:

Example 1. Redefine $tan(240°)$ to an acute angle.
Solution: Using $tan(x) = tan(k\pi + x)$: $tan(240°) = tan(180° + 60°) = tan(60°)$.

Example 2. Solve $(sin(210°) + sin(150°) + cot(210°) + cot(330°))$.
Solution: We find each one using the formulas $sin(x) = sin(2k\pi + x)$ and $cot(x) = cot(k\pi + x)$, then combine all in the end, so:
$sin(210°) = sin(180° + 30°) = -sin(30°)$, $sin(150°) = sin(180° - 30°) = sin(30°)$.
$cot(210°) = cot(180° + 30°) = cot(30°)$, $cot(330°) = cot(360° + (-30°)) = -cot(30°)$. By combining: $sin(210°) + sin(150°) + cot(210°) + cot(330°) = -sin(30°) + sin(30°) + cot(30°) - cot(30°) = 0$.

bit.ly/3qZbj1x

Floor and Ceiling Functions

- You know of the absolute value function and its symbol: "$|x|$"

  $|-3.2| = 3.2$ and $|4.1| = 4.1$

  There are other operators that look similar but serve different purposes.

- **Floor function** of a real number, returns the largest integer that is less than or equal to the variable inside it. It is denoted using $\lfloor x \rfloor$ and basically rounds the x down to the nearest integer.

  For example: $\lfloor 3.7 \rfloor = 3$.

- Floor function is also called "the greatest integer function", and is also shown using the brackets symbol: $[x]$.

- **Ceiling function** of a real number, returns the smallest integer that is greater than or equal to the variable inside it. It is denoted using $\lceil x \rceil$ and basically rounds the x up to the nearest integer. For example: $\lceil 3.7 \rceil = 4$.

- Here are some more examples for these functions:

  $\lfloor 4.3 \rfloor = 4$ $\lfloor -2.8 \rfloor = -3$ $\lfloor 0.999 \rfloor = 0$ $\lfloor 7 \rfloor = 7$

  $\lceil 4.3 \rceil = 5$ $\lceil -2.8 \rceil = -2$ $\lceil 0.001 \rceil = 1$ $\lceil 7 \rceil = 7$

Examples:

Example 1. Solve this equation: $|-4.36| + \lceil 3.2 \rceil - \lfloor -1.5 \rfloor$.

Solution: Absolute value of $-4.36 = 4.36$, ceiling of 3.2 is 4, and floor of -1.5 is -2, so we have: $4.36 + 4 - (-2) = 10.36$.

Example 2. Simplify the equation:

$$|sin(270°)| + \lceil cos(45°) \rceil - \lceil tan(60°) \rceil + \lfloor cos(30°) \rfloor$$

Solution: We go one by one, then combine the results: $|sin(270°)| = |-1| = 1$,

$\lceil cos(45°) \rceil = \lceil \frac{\sqrt{2}}{2} \rceil = \lceil \frac{1.41}{2} \rceil = \lceil 0.7 \rceil = 1$, $\lceil tan\, 60° \rceil = \lceil \sqrt{3} \rceil = \lceil 1.73 \rceil = 2$,

$\lfloor cos(30°) \rfloor = \lfloor \frac{\sqrt{3}}{2} \rfloor = \lfloor \frac{1.73}{2} \rfloor = \lfloor 0.86 \rfloor = 0$.

So, when combined, we have: $1 + 1 - 2 + 0 = 0$.

Please note that for floor/ceiling functions, we only need to deal with integers and can find the approximate answer for the radical instead of the exact value and calculate the floor/ceiling using our educated guess.

Chapter 2: Practices

✎ **Determine whether the following table of values represents a linear function, an exponential function, or a quadratic function.**

1) _____

x	-2	-1	0	1	2
y	$\frac{1}{2}$	1	2	4	8

2) _____

x	-2	-1	0	1	2
y	5	2	1	2	5

3) _____

x	-2	-1	0	1	2
y	-4	-1	2	5	8

✎ **Expand each logarithm.**

4) $\log_b(2 \times 9) =$ ____

5) $\log_b\left(\frac{5}{7}\right) =$ ____

6) $\log_b(xy) =$ ____

7) $\log_b(2x^2 \times 3y) =$ ____

✎ **Reduce the following expressions to simplest form.**

8) $e^{\ln 4 + \ln 5} =$ ____

9) $e^{\ln\left(\frac{9}{e}\right)} =$ ____

10) $e^{\ln 2 + \ln 7} =$ ____

11) $6\ln(e^5) =$ ____

Find the value of the variables in each equation.

12) $\log_3 8x = 3 \Rightarrow x =$ _____

13) $\log_4 2x = 5 \Rightarrow x =$ _____

14) $\log_4 5x = 0 \Rightarrow x =$ _____

15) $\log 4x = \log 5 \Rightarrow x =$ _____

Evaluate.

16) $\sin(120°) =$

17) $\sin(-330°) =$

18) $\tan(-90°) =$

19) $\cot(90°) =$

20) $\cos(-90°) =$

21) $\sec(60°) =$

22) $\csc(480°) =$

23) $\cot(-135°) =$

Determine the following trigonometric functions using values from the triangle:

24) $\sin(\theta) =$ _____

25) $\cot(\beta) =$ _____

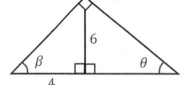

Simplify the given trigonometric expressions.

26) $\dfrac{1-\sin^2(x)}{\frac{1}{\sec(x)} \cdot \tan(x)} + \sin(x) =$ _____

27) $(\tan(x) + \cot(x)) \times \dfrac{\sec(x)}{\csc(x)} =$ _____

Find the domain and range of functions.

28) $y = \cos(x) - 4$

Domain: _____

Range: _____

29) $y = \sin(x) - 3$

Domain: _____

Range: _____

30) $y = \dfrac{1}{2-\sin(2x)}$

Domain: _____

Range: _____

31) $y = 2\cos(x) + 6$

Domain: _____

Range: _____

Chapter 2: Answers

1) Exponential   2) Quadratic   3) Linear

4) $\log_b 2 + 2\log_b 3$ 6) $\log_b x + \log_b y$

5) $\log_b 5 - \log_b 7$ 7) $2\log_b 2 + 2\log_b x + \log_b 3 + \log_b y$

8) 20   9) $\frac{9}{e}$ 10) 14   11) 30

12) $\frac{27}{8}$ 13) 512   14) $\frac{1}{5}$ 15) $\frac{5}{4}$

16) $\frac{\sqrt{3}}{2}$ 20) 0

17) $\frac{1}{2}$ 21) 2

18) Undefined   22) $\frac{2\sqrt{3}}{3}$

19) 0   23) 1

24) $\frac{\sqrt{52}}{13}$ 25) $\frac{2}{3}$

26) $\csc(x)$ 27) $\sec^2(x)$

28) Domain: $(-\infty, +\infty)$ 30) Domain: $(-\infty, +\infty)$

Range: $[-5, -3]$ Range: $\left[\frac{1}{3}, 1\right]$

29) Domain: $(-\infty, +\infty)$ 31) Domain: $(-\infty, +\infty)$

Range: $[-4, -2]$ Range: $[4, 8]$

CHAPTER
3 Sequences and Series

Math topics that you'll learn in this chapter:

- ☑ Arithmetic Sequences
- ☑ Geometric Sequences
- ☑ Sigma Notation (Summation Notation)
- ☑ Arithmetic Series
- ☑ Geometric Series
- ☑ Binomial Theorem
- ☑ Pascal's Triangle
- ☑ Alternate Series

Arithmetic Sequences

- A sequence of numbers such that the difference between the consecutive terms is constant, is called an arithmetic sequence. For example, the sequence 6, 8, 10, 12, 14, ⋯ is an arithmetic sequence with common difference of 2.

- To find any term in an arithmetic sequence use this formula:

  $a_n = a_1 + (n-1)d$, where $a_1 =$ the first term, $d =$ the common difference between terms, $n =$ term number

- Common difference is found by subtracting two consecutive terms, or:

  $$a_{n+1} - a_n = d$$

  It can also be calculated using this formula: $\frac{a_m - a_n}{m - n}$.

Examples:

Example 1. Find the first five terms of the sequence. $a_8 = 38, d = 3$

Solution: First, we need to find a_1 or a. Use the arithmetic sequence formula: $a_n = a_1 + (n-1)d$. If $a_8 = 38$, then $n = 8$. Rewrite the formula and put the values provided: $a_n = a + d(n-1) \Rightarrow 38 = a + 3(8-1) = a + 21$.

Now solve for a: $38 = a + 21 \Rightarrow a = 38 - 21 = 17$.

First five terms: 17, 20, 23, 26, 29.

Example 2. Given the first term and the common difference of an arithmetic sequence find the first five terms. $a_1 = 18, d = 2$.

Solution: Use the arithmetic sequence formula: $a_n = a_1 + (n-1)d$.

First five terms: 18, 20, 22, 24, 26.

Example 3. Find the missing terms of this arithmetic sequence:

$$\{a_1, a_2, a_3, a_4, -4, 5, a_7\}$$

Solution: We have 6th and 7th term. Since we know that this is an arithmetic sequence, common difference is: $d = 5 - (-4) = 9$, and we know that $a_6 = a + 5d$, so: $5 = a + 5(9) \Rightarrow a = -40$, so other terms are:

$\{-40, -31, -22, -13, -4, 5, 14\}$.

Geometric Sequences

- It is a sequence of numbers where each term after the first is found by multiplying the previous item by the common ratio, a fixed, non-zero number. For example, the sequence 2, 4, 8, 16, 32, ⋯ is a geometric sequence with a common ratio of 2.

- To find any term in a geometric sequence use this formula: $a_n = ar^{(n-1)}$.

  a = the first term, r = the common ratio, n = term number.

- Common ratio is found by dividing two consecutive terms, or: $r = \frac{a_{n+1}}{a_n}$.

  It can also be calculated using: $r^{m-n} = \frac{a_m}{a_n}$.

Examples:

Example 1. Given the first term and the common ratio of a geometric sequence find the first five terms of the sequence. $a_1 = 3, r = -2$

Solution: Use geometric sequence formula: $a_n = ar^{(n-1)} \Rightarrow a_n = 3(-2)^{n-1}$.

$a_2 = ar = 3(-2) = -6$, $a_3 = ar^2 = 3(-2)^2 = 3(4) = 12$, $a_4 = ar^3 = 3(-8) = -24$, and $a_5 = ar^4 = 3(16) = 48$, so overall: $\{3, -6, 12, -24, 48, \cdots\}$.

Example 2. Given two terms in a geometric sequence find the 8th term. $a_3 = 10$, and $a_5 = 40$.

Solution: Use geometric sequence formula: $a_n = ar^{(n-1)}$.

So: $a_3 = ar^{(3-1)} = ar^2 = 10$, and: $a_5 = ar^{(5-1)} = ar^4 = 40$.

Now divide a_5 by a_3. Then: $\frac{a_5}{a_3} = \frac{ar^4}{ar^2} = r^2 \Rightarrow r^2 = \frac{40}{10} = 4 \Rightarrow r = 2$.

We can find a now: $ar^2 = 10 \Rightarrow a(2^2) = 10 \Rightarrow a = 2.5$.

Use the formula to find the 8th term:

$a_n = ar^{(n-1)} \Rightarrow a_8 = (2.5)(2)^{(8-1)} = 2.5(2)^7 = 2.5(128) = 320$.

Example 3. Find the missing terms of this geometric sequence:

$\{a_1, a_2, 5, a_4, a_5, 0.625\}$.

Solution: We use the formula $r^{m-n} = \frac{a_m}{a_n}$, so: $r^{6-3} = \frac{0.625}{5} = 0.125 \Rightarrow r^3 = 0.125$.

So: $r = 0.5$, and if $r = 0.5$: $a_3 = ar^2 \Rightarrow 5 = a(0.5)^2$.

So $a = \frac{5}{0.25} = 20$, so the missing terms are: $\{20, 10, 5, 2.5, 1.25, 0.625\}$.

Sigma Notation (Summation Notation)

- Summation notation is a way to express the sum of terms of a sequence in abbreviated form.

- Let $a_1, a_2, \cdots, a_i, \cdots$ be a sequence with the starting term a_1 and the ith term a_i. The representation of the sum of the kth term to the nth term of this sequence is as follows:

$$\sum_{i=k}^{n} a_i$$

Stopping point (top, n), Summation sign Called sigma (Σ), index (i), Starting point ($i=k$), Formula, function of i (a_i).

- The sigma (Σ) operator represents the sum of a sequence, specifying the terms to be summed and their range.

- For sequences a_i and b_i and positive integer n:

  - $\sum_{i=1}^{n} ca_i = c \sum_{i=1}^{n} a_i$
  - $\sum_{i=1}^{n} (a_i + b_i) = \sum_{i=1}^{n} a_i + \sum_{i=1}^{n} b_i$
  - $\sum_{i=1}^{n} c = nc$
  - $\sum_{i=1}^{n} i = \frac{n(n+1)}{2}$
  - $\sum_{i=1}^{n} i^2 = \frac{n(n+1)(2n+1)}{6}$

Example:

Evaluate: $\sum_{k=1}^{10} (k^2 - 1)$.

Solution: Using this property $\sum_{i=1}^{n} (a_i + b_i) = \sum_{i=1}^{n} a_i + \sum_{i=1}^{n} b_i$, we have:

$\sum_{k=1}^{10} (k^2 - 1) = \sum_{k=1}^{10} k^2 - \sum_{k=1}^{10} 1$. Now, use these formulas $\sum_{i=1}^{n} i^2 = \frac{n(n+1)(2n+1)}{6}$ and $\sum_{i=1}^{n} c = nc$, so: $\sum_{k=1}^{10} 1 = 10$, and $\sum_{k=1}^{10} k^2 = \frac{10(10+1)(2\times 10+1)}{6} = 385$.

Therefore: $\sum_{k=1}^{10} (k^2 - 1) = 385 - 10 = 375$.

Arithmetic Series

- An arithmetic series is the sum of sequence in which each term is computed from the previous one by adding (or subtracting) a constant value d.
- The sum of the sequence of the first n terms is given by:

$$S_n = \sum_{k=1}^{n} a_k = \sum_{k=1}^{n} [a_1 + (k-1)d] = na_1 + d\sum_{k=1}^{n}(k-1)$$

- We can show the sum of first n term with:

$$S_n = a + (a+d) + (a+2d) + \cdots + (a+(n-2)d) + (a+(n-1)d)$$

Let's write it in reverse:

$$S_n = (a+(n-1)d) + (a+(n-2)d) + \cdots + (a+2d) + (a+d) + a$$

If we add corresponding terms of S_n and its inverse, we end up with:

$$2S_n = (2a+(n-1)d) + (2a+(n-1)d) + \cdots + (2a+(n-1)d)$$

So $2S_n$ is equals n times the $(2a+(n-1)d)$: $2S_n = n(2a+(n-1)d)$

$S_n = \frac{n}{2}(a + a + (n-1)d)$, so: $S_n = \frac{1}{2}n(a_1 + a_n)$.

This is called "Gauss method" for calculating a sum. We can also use the sum identify $\sum_{i=1}^{n} i = \frac{n(n+1)}{2}$: $S_n = na_1 + \frac{1}{2}dn(n-1) = \frac{1}{2}n[2a_1 + d(n-1)] = \frac{1}{2}n[a_1 + a_1 + d(n-1)] = \frac{1}{2}n[a_1 + a_n]$.

Examples:

Example 1. In the arithmetic series $4, 11, 18, \cdots$, find the sum of the first 10 terms.

Solution: $a_1 = 4$, $d = 11 - 4 = 7$, $n = 10$. Use the arithmetic series formula to find the sum: $S_n = \frac{1}{2}n[2a_1 + d(n-1)]$. Therefore:

$S_{10} = \frac{10}{2}[2(4) + 7(10-1)] \Rightarrow S_{10} = 5(8 + 63) \Rightarrow S_{10} = 355$.

Example 2. Find the sum of the first 4 terms of the sequence. $a_{10} = 46$, $d = 4$

Solution: First, we need to find a_1 or a. Use the arithmetic sequence formula:

$a_n = a + d(n-1) \Rightarrow 46 = a + 4(10-1)$, now solve for a.

$$46 = a + 36 \Rightarrow a = 46 - 36 = 10$$

First four terms: $10, 14, 18, 22$. Therefore: $10 + 14 + 18 + 22 = 64$, or:

$S_n = \frac{n}{2}[2a_1 + d(n-1)] = \frac{4}{2}[2(10) + 4(4-1)] = 64$.

Geometric Series

- A geometric sequence involves a common ratio, and this ratio can have an absolute value ($|r|$) either greater than, equal to, or less than 1.
- If $|r| < 1$, the series converges, and the sum of all its terms can be calculated using the formula: Infinite geometric series: $S = \sum_{i=0}^{\infty} a_i r^i = \frac{a_1}{1-r}$.
- If $|r| = 1$, the series does not converge to a finite sum unless all terms are zero after a certain point.
- If $|r| > 1$, the series diverges, and the sum of all its terms is infinite. However, you can calculate the sum of the **first n terms** using the formula for the partial sum of a geometric series:

    Finite geometric sequence: $S_n = \sum_{i=1}^{n} a_1 r^{i-1} = a_1 \left(\frac{1-r^n}{1-r}\right) = a_1 \left(\frac{r^n-1}{r-1}\right)$.

Examples:

Example 1. $\sum_{n=1}^{4} 3^{n-1}$.

Solution: Use this formula: $S_n = \sum_{i=1}^{n} a_1 r^{i-1} = a_1 \left(\frac{1-r^n}{1-r}\right) \Rightarrow \sum_{n=1}^{4} 3^{n-1} = (1)\left(\frac{1-3^4}{1-3}\right)$.
Then: $(1)\left(\frac{1-3^4}{1-3}\right) = 1\left(\frac{1-81}{1-3}\right) = \left(\frac{-80}{-2}\right) = 40$.

Example 2. $\sum_{n=1}^{5} -2^{n-1}$.

Solution: $\sum_{n=1}^{5} -2^{n-1} = (-1)\left(\frac{1-2^5}{1-2}\right)$. Then: $(-1)\left(\frac{1-32}{1-2}\right) = (-1)\left(\frac{-31}{-1}\right) = -31$.

Example 3. Evaluate the geometric series described. $\sum_{i=1}^{\infty} \left(\frac{1}{3}\right)^{i-1}$

Solution: The absolute value of the ratio is $\frac{1}{3}$. Using formula:

$\sum_{i=0}^{\infty} a_i r^i = \frac{a_1}{1-r} \Rightarrow \sum_{i=1}^{\infty} \left(\frac{1}{3}\right)^{i-1} = \frac{1}{1-\frac{1}{3}} = \frac{1}{\frac{2}{3}} = \frac{3}{2}$.

Example 4. Evaluate the geometric series described.

$\sum_{k=1}^{\infty} \left(\frac{1}{4}\right) 7^{k-1}$

Solution: Since the absolute value of the ratio is 7 and more than 1, the geometric series is infinite.

Term	Value
1	1
2	0.333333333333333000
3	0.111111111111111000
4	0.037037037037037000
5	0.012345679012345700
6	0.004115226337448560
7	0.001371742112482850
8	0.000457247370827618
9	0.000152415790275873
10	0.000050805263425291
11	0.000016935087808430

bit.ly/3IN3YsG
Find more at

Binomial Theorem

- The binomial theorem states that $(x + y)^n$, expands into a sum of terms, each with a binomial coefficient, x^{n-k} and y^k, where k ranges from 0 to n.

- The formula of the binomial theorem for positive integer n is as follows:

$$(x + y)^n = \sum_{k=0}^{n} \binom{n}{k} x^{n-k} y^k$$

$$= \binom{n}{0} x^n y^0 + \binom{n}{1} x^{n-1} y^1 + \cdots + \binom{n}{n-1} x^1 y^{n-1} + \binom{n}{n} x^0 y^n$$

Where $\binom{n}{k} = \frac{n!}{k!(n-k)!}$, and $0 \leq k \leq n$.

- In the expansion of $(x + y)^n$, there are $n + 1$ terms and the kth term is equal to: $\binom{n}{k-1} x^{n-k+1} y^{k-1}$.

Examples:

Example 1. Write the expansion $(x + y)^4$.

Solution: $(x + y)^4 = \sum_{k=0}^{4} \binom{4}{k} x^{4-k} y^k = \binom{4}{0} x^4 + \binom{4}{1} x^3 y + \binom{4}{2} x^2 y^2 + \binom{4}{3} xy^3 + \binom{4}{4} y^4 = x^4 + 4x^3 y + 6x^2 y^2 + 4xy^3 + y^4$.

Example 2. Write the 3rd term of the expansion of $(a - 1)^5$.

Solution: Use this formula: The kth term $= \binom{n}{k-1} x^{n-k+1} y^{k-1}$.

The 3rd term $= \binom{5}{3-1}(a)^{5-3+1}(-1)^{3-1} = \binom{5}{2}(a)^3 (-1)^2 = 10a^3$.

Example 3. Write the expansion $(2b + 2)^3$.

Solution: $(2b + 2)^3 = \sum_{k=0}^{3} \binom{3}{k}(2b)^{3-k}(2)^k = \binom{3}{0}(2b)^3 + \binom{3}{1}(2b)^2 (2) + \binom{3}{2}(2b)(2)^2 + \binom{3}{3}(2)^3 = 8b^3 + 24b^2 + 24b + 8$.

Example 4. Write the 5th term of the expansion $(-a + 2b)^6$.

Solution: $\binom{n}{k-1} x^{n-k+1} y^{k-1} = \binom{6}{4}(-a)^2 (2b)^4 = \frac{6!}{4!2!} a^2 \times 16b^4 = 15a^2 \times 16b^4 = 240a^2 b^4$.

Pascal's Triangle

- Pascal's triangle is shown below:

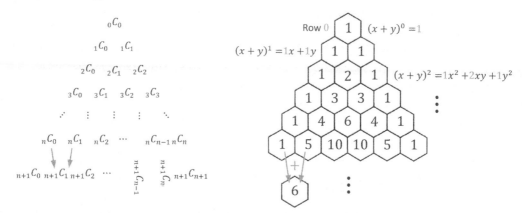

- The nth row of Pascal's triangle contains $n + 1$ components.

- $_nC_{k-1} + {_nC_k} = {_{n+1}C_k}$, where $0 < k \leq n$.

- In the nth row of Pascal's triangle, the kth component is equal to $_nC_{k-1}$.

- For all entries in Pascal's triangle in row n: $\sum_{k=0}^{n} {_nC_k} = 2^n$.

- Each component, has a maximum exponent degree of n, for example, at $n = 2$, each of $1x^2$, $2xy$ and y^2, are of 2nd degree.

Examples:

Example 1. Find the 5th entry in row 7 of Pascal's triangle.

Solution: The kth entry in row n of Pascal's triangle is $_nC_{k-1}$.

(5th entry in row 7) $= {_7C_{5-1}} = {_7C_4} = \frac{7!}{3!4!} = \frac{7 \times 6 \times 5 \times 4!}{3 \times 2 \times 4!} = 35$.

Example 2. Find the 8th entry in row 10 of Pascal's triangle.

Solution: The kth entry in row n of Pascal's triangle is $_nC_{k-1}$.

(8th entry in row 10) $= {_{10}C_{8-1}} = {_{10}C_7} = \frac{10!}{7!3!} = 120$.

Example 3. Find the location of $_{10}C_2$ entry in Pascal's triangle. Then give its value of it.

Solution: We know that the kth entry in row n of Pascal's triangle is $_nC_{k-1}$. Then the value of n and k for $_{10}C_2$ is $n = 10$ and $k - 1 = 2 \Rightarrow k = 3$. So $_{10}C_2$ is the 3rd entry in row 10 of Pascal's triangle. Now, $_{10}C_2 = \frac{10!}{2!8!} = 45$.

Alternate Series

- An alternate series is a mathematical series with terms that alternate between positive and negative values, often studied for convergence or divergence.

- The general form of an alternating series is as follows:

$$\sum_{i=1}^{\infty}(-1)^k a_k$$

Where $a_k \geq 0$ and the first index is arbitrary. It means that the starting term for an alternating series can have any sign.

- An alternating series $\{a_k\}_{k=1}^{\infty}$ is called **convergent** if the absolute value of its terms keeps getting smaller as you go further in the series, and as you look at more and more terms in the series, those values become extremely close to zero:

  - $0 \leq a_{k+1} \leq a_k$, for all $k \geq 1$.
  - $a_k \to 0$, as $k \to +\infty$.

Otherwise, the series is **divergent**.

Examples:

Example 1. Determine whether the following series converge or diverge:

$$\sum_{i=1}^{\infty}(-1)^i \frac{2}{i+5}$$

Solution: Let $a_i = \frac{2}{i+5}$. Then: $\frac{2}{i+5} \to 0$ as $i \to +\infty$. In addition, $0 \leq a_{k+1} \leq a_k$:

$\frac{2}{(i+1)+5} \leq \frac{2}{i+5} \Rightarrow \frac{2}{i+6} \leq \frac{2}{i+5} \Rightarrow i+6 \geq i+5 \Rightarrow 6 \geq 5$.

This is true. Therefore, the alternating series of the problem is convergent.

Example 2. Determine whether the following series converge or diverge:

$$\sum_{k=1}^{\infty}\frac{(-1)^k k}{2k+1}$$

Solution: Let $a_k = \frac{k}{2k+1}$. Then: $\frac{k}{2k+1} \to \frac{1}{2} \neq 0$ as $i \to +\infty$.

Therefore, the alternating series of the problem is divergent.

Chapter 3: Practices

Find the next three terms of each arithmetic sequence.

1) 15, 11, 7, 3, −1, __, __, __,

2) −21, −14, −7, 0, __, __, __,

3) 3, 6, 9, 12, 15, __, __, __,

4) 4, 8, 12, 16, 20, __, __, __,

Given the first term and the common difference of an arithmetic sequence find the first five terms and the explicit formula.

5) $a_1 = 24, d = 2$
__, __, __, __, __, ⋯, $a_n =$

6) $a_1 = -15, d = -5$
__, __, __, __, __, ⋯, $a_n =$

7) $a_1 = 18, d = 10$
__, __, __, __, __, ⋯, $a_n =$

8) $a_1 = -38, d = -10$
__, __, __, __, __, ⋯, $a_n =$

Find the first five terms of the sequence.

9) $a_1 = -120, d = -100$ __, __, __, __, __, ⋯

10) $a_1 = 55, d = 23$ __, __, __, __, __, ⋯

11) $a_1 = 12.5, d = 4.2$ __, __, __, __, __, ⋯

Determine if the sequence is geometric. If it is, find the common ratio.

12) 1, −5, 25, −125, ⋯; $r =$ __

13) −2, −4, −8, −16, ⋯; $r =$ __

14) 4, 16, 36, 64, ⋯; $r =$ __

15) −3, −15, −75, −375, ⋯; $r =$ __

Given the first term and the common ratio of a geometric sequence find the first five terms and the explicit formula.

16) $a_1 = 0.8, r = -5$ __, __, __, __, __, ⋯, $a_n =$ __

17) $a_1 = 1, r = 2$ __, __, __, __, __, ⋯, $a_n =$ __

Effortless Math Education

✎ Evaluate each geometric series described.

18) $1, +2, +4, +8, \cdots, n = 6$

19) $1, -4, +16, -64, \cdots, n = 9$

20) $-2, -6, -18, -54, \cdots, n = 9$

21) $2, -10, +50, -250, \cdots, n = 8$

22) $1, -5, +25, -125, \cdots, n = 7$

23) $-3, -6, -12, -24, \cdots, n = 9$

✎ Determine if each geometric series converges or diverges.

24) $a_1 = -1, r = 3$

25) $a_1 = 3.2, r = 0.2$

26) $a_1 = 5, r = 2$

27) $-1, 3, -9, 27, \cdots$

28) $2, -1, \frac{1}{2}, -\frac{1}{4}, \frac{1}{8}, \cdots$

29) $81, 27, 9, 3, \cdots$

✎ Solve.

30) Find the 6th entry in row 8 of Pascal's triangle. _____

31) Find the 5th entry in row 9 of Pascal's triangle. _____

✎ Solve.

32) Write the 5th term of the expansion of $(1 - 4b^2)^4$. _____

33) Write the expansion $(2x^2 - 5)^3$. _____

✎ Write the terms of the series.

34) $\sum_{j=1}^{6} 4(j-2)^2 =$ _____

35) $\sum_{b=1}^{5} (b^2 - 4)^2 =$ _____

36) $\sum_{k=3}^{10} 2(k+3) =$ _____

37) $\sum_{x=1}^{8} (3x^2 + 2) =$ _____

Determine whether the following series converge or diverge.

38) $\sum_{n=1}^{\infty} \frac{(-1)^n}{n^2} = $ _____

39) $\sum_{i=1}^{\infty} (3)^i \frac{1}{i-2} = $ _____

40) $\sum_{n=1}^{\infty} \frac{n^2+1}{n^3+1} = $ _____

41) $\sum_{i=1}^{\infty} \frac{(-1)^{i+3}}{i^2+4i+2} = $ _____

Chapter 3: Answers

1) $-5, -9, -13$

2) $7, 14, 21$

3) $18, 21, 24$

4) $24, 28, 32$

5) First Five Terms: $24, 26, 28, 30, 32$, Explicit: $a_n = 2n + 22$

6) First Five Terms: $-15, -20, -25, -30, -35$, Explicit: $a_n = -5n - 10$

7) First Five Terms: $18, 28, 38, 48, 58$, Explicit: $a_n = 10n + 8$

8) First Five Terms: $-38, -48, -58, -68, -78$, Explicit: $a_n = -10n - 28$

9) $-120, -220, -320, -420, -520$

10) $55, 78, 101, 124, 147$

11) $12.5, 16.7, 20.9, 25.1, 29.3$

12) $r = -5$

13) $r = 2$

14) Not geometric

15) $r = 5$

16) First Five Terms:
    $0.8, -4, 20, -100, 500$
    $a_n = 0.8(-5)^{n-1}$

17) First Five Terms: $1, 2, 4, 8, 16$
    $a_n = 2^{n-1}$

18) 63

19) $52,429$

20) $-19,682$

21) $-130,208$

22) $13,021$

23) $-1,533$

24) Diverges

25) Converges

26) Diverges

27) Diverges

28) Converges

29) Converges

30) 56

31) 126

32) $256b^8$

33) $8x^6 - 60x^4 + 150x^2 - 125$

34) 124

35) 619

36) 152

37) 628

38) Converge

39) Diverge

40) Diverge

41) Converge

CHAPTER 4
Limit and Continuity

Math topics that you'll learn in this chapter:

- ☑ Limit Introduction
- ☑ Neighborhood
- ☑ Estimating Limits from Tables
- ☑ Functions with Undefined Limits
- ☑ One Sided Limits
- ☑ Limit at Infinity
- ☑ Continuity at a point and over an interval
- ☑ Removing Discontinuity
- ☑ Direct Substitution
- ☑ Limit Laws
- ☑ The Squeeze Theorem
- ☑ Indeterminates and undefined
- ☑ Infinity cases
- ☑ Trigonometric limits and rationalizing
- ☑ Algebraic manipulation
- ☑ Redefine function's value.
- ☑ Rationalizing Infinite limits

Limit Introduction

- A limit looks at what happens to a function when the input approaches a certain value.

- The general notation for a limit is as follows: $\lim\limits_{x \to a} f(x)$

  which is read as "The limit of f of x as x approaches a"

- Suppose function f is defined for x near a. (and not necessarily at $x = a$ itself)

  We define L as the limit of $f(x)$ as x approaches a, meaning as x approaches a, limit approaches L: $\lim\limits_{x \to a} f(x) = L$.

- If ∞ is the answer to a limit, we say the limit doesn't have an answer.

- We can estimate a limit from the graph of the function, by finding the value that limit approaches (a), on x−axis, draw a line upwards (or downwards) until the line hits the graph, and find the corresponding value on y−axis (perpendicular). If by closing in towards a from both sides, we get close to a particular value on y axis, that value is our limit.

- So, if you approach a point (on x−axis) from both the left and right sides on a graph, and the y−values you get are the same, then the limit exists.

Examples:

Example 1. $\lim\limits_{x \to 2} 2x - 3$.

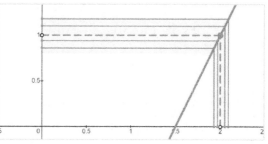

Solution: As you can see from the graph of this function, the closer we get to $x = 2$, the $f(x)$ or y gets closer and closer to 1. That's our limit.

If x were to approach 1.5 ($x \to 1.5$) the limit would approach 0. This is the graph of $\frac{2}{13}x^2 - \frac{7}{3}x$, find the limit as $x \to 15$.

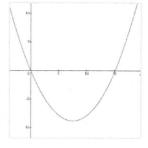

Solution: If we approach 15 from both sides, the y value approaches 0 and that's our limit.

Neighborhood

- On the real number line, numbers within a certain distance are called "neighborhood" of that number.

- Given a point, calling it "a" and ε being a positive real number representing the size of the interval, a <u>right neighborhood</u> of "a" consists of all the numbers that are greater than "a" and are within ε distance to the right of "a". Symbolically, a right neighborhood of "a" can be represented as $(a, a + \varepsilon)$, which means a itself is not part of the interval, or $[a, a + \varepsilon)$ which means a itself is part of the interval.

- Similarly, a <u>left neighborhood</u> of "a" consists of all the numbers that are less than "a" and are within ε distance to the left of "a" and a left neighborhood of "a" can be represented as $(a - \varepsilon, a)$ or $(a - \varepsilon, a]$.

- So general neighborhood of a can be mathematically shown as: $(a - \varepsilon, a + \varepsilon)$

  neighborhood

  $a-\varepsilon \quad a \quad a+\varepsilon$

- It is required for a function to be defined in the left neighborhood of a if we were to approach a with x values smaller than a (and vice versa), and not necessarily at a itself, because as we approach a, we don't actually need to get to it.

- "Deleted neighborhood" or "Deleted open interval" of a, refers to the neighborhood or interval that excludes the point a.

- A neighborhood of a can also be shown as: $\{x : |x - a| < \varepsilon\}$.

- As a general rule, an interval like (α, β) is a neighborhood of the point $\frac{\alpha+\beta}{2}$, which is the intervals midpoint.

Examples:

Example 1. What point is the interval $\left(\frac{1}{2}, \frac{3}{4}\right)$ a neighborhood of?

Solution: $\frac{1}{2} + \frac{3}{4} = \frac{10}{8} \div 2 = \frac{10}{16} = \frac{5}{8}$

Example 2. Find the midpoint of the neighborhood that surrounds the interval $(-2.5, 1.8)$.

Solution: Using the formula, $\frac{\alpha+\beta}{2}$: $\frac{-2.5+1.8}{2} = -0.35$

Estimating Limits from Tables

- One method of evaluating limits is by choosing inputs close to the point of interest (from both sides) and observing the behavior of the function's output.

- It relies on observed values rather than a strict mathematical proof, so it can give a good approximation or insight into what the limit might be, but it doesn't always provide a conclusive or exact answer. For example, some functions have complex behavior near the limit point, or they might oscillate rapidly. In such cases, estimating using the tables may not accurately capture the function's behavior.

Examples:

Example 1. $\lim\limits_{x \to 3} 2x$.

Solution: We approach from left:

x	2.8	2.85	2.9	2.99
$2x$	5.6	5.7	5.8	5.98

And right:

x	3.3	3.2	3.1	3.01
$2x$	6.6	6.4	6.2	6.02

As you can see, as we get closer and closer to 3, in both cases the limit is approaching number 6. So, we can confidently say that: $\lim\limits_{x \to 3} 2x = 6$.

Example 2. Find the limit: $\lim\limits_{x \to 2} \frac{x-2}{x^2-2x}$.

Solution: We need to get close to number 2 from left and right. From left, we choose {1.9,1.99,1.999}, which if placed in the function, outputs {0.5263,0.5025,0.5002}, and if we approach 2 from right, let's say {2.1,2.01,2.001}, we get {0.4762,0.4975,0.4998}. So, limit as $x \to 2$, exists and is 0.5.

Note that we cannot have zero in the denominator, so for $x = \{0,2\}$, the function is undefined.

Functions with Undefined Limits (from table)

- Not every function has a limit. We can use tables and graphs to identify these kinds of functions.

- If by approaching a value from left and right, we get two different limits, no particular limit, or infinity, we say "the limit is undefined for that value".

Examples:

Example 1. $\lim\limits_{x \to 2} \frac{1}{x-2}$.

Solution: We approach 2 from both sides and substitute values closer and closer to 2 in the function:

1.89	1.9	1.99	1.999	2	2.001	2.01	2.1	2.11
−9.09	−10	−100	−1000	undefined	1000	100	10	9.09

As you can see, by approaching 2, the function is not getting close to a particular number, so it's limit is undefined.

Example 2. $\lim\limits_{x \to 0} \sin\frac{1}{x}$.

Solution: From left and right, we can have:

−0.1	−0.01	−0.001	−1.999	0	0.0001	0.001	0.01	0.1
0.544	0.506	−0.826	−0.479	undefined	−0.305	0.826	−0.506	−0.544

As you can see, this function oscillates between 1 and −1 as we get closer to zero. Since we are not approaching a particular value, we don't have a limit at this point. (You can see the graph of this function in the next page)

Functions with Undefined Limits (from graphs)

We can determine if a function can have a limit or not, from its graph. These are just a few examples, and there are other types of functions that may exhibit behaviors where limits do not exist (in a point).

1. Oscillating functions, that oscillate infinitely between different values as the input approaches a particular point. (Graph of $\lim\limits_{x \to 0} \sin \frac{1}{x}$)

2. Functions that approaach different values. consider $f(x) = \frac{1}{x}$. As x approaches 0, $f(x)$ approaches $+\infty$ or $-\infty$, depending on whether x approaches 0 from the positive or negative side. Since the function does not approach a specific value, the limit does not exist at $x = 0$.

   Or Functions that behave differently at left and right of $x = a$, don't have a limit at a.

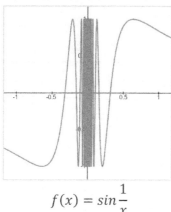
$f(x) = \sin \frac{1}{x}$

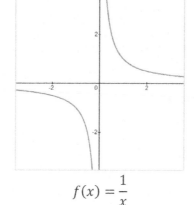
$f(x) = \frac{1}{x}$

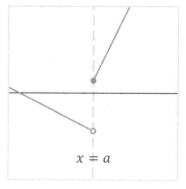
$x = a$

3. Divergent functions: Functions that grow without bound as the input goes to ∞. A typical example is the exponential function, $f(x) = e^x$. As $x \to \infty$, the function increases exponentially, and there is no finite value to which it converges. Hence, the limit does not exist.

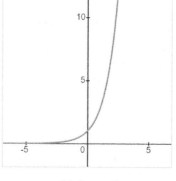
$f(x) = e^x$

One Sided Limits

- One-sided limits are used to analyze the behavior of a function as it approaches a specific point from either the left or the right side.

- This can be helpful when dealing with functions that have discontinuities, sharp turns, or asymptotic behavior.

- One-sided limits are particularly useful in determining the continuity of a function.

- If function of $f(x)$ gets closer and closer to L when we approach a with x values greater but sufficiently close to it, we call L the one-sided limit of $f(x)$ and we show it as: $\lim\limits_{x \to a^+} f(x) = L$.

  Similarly, if we approach a from the left, we show the one-sided limit as: $\lim\limits_{x \to a^-} f(x) = L$.

- To discuss whether a function has right limit at point a, it must be defined at the right neighborhood of a (same goes for left neighborhood)

- Note that for the limit of a function to exist at a point, both left and right limits must exist and be equal. But it is a common <u>convention</u> in mathematics that if a function has a limit only from one side (either the left side or the right side) at a particular point, then the limit of the function at that point is equal to the limit from that specific side. Denoted as: $\lim\limits_{x \to a} f(x) = \lim\limits_{x \to a^+} f(x)$ or $\lim\limits_{x \to a} f(x) = \lim\limits_{x \to a^-} f(x)$.

  This convention holds as long as the function has a limit from at least one side and the other side does not produce a contradictory or different limit.

Example:

Example 1. Solve: $\lim\limits_{x \to 3} \sqrt{3 - x}$.

Solution: This function is undefined for $x > 3$, so if we were to find the limit, we couldn't approach 3 from the right side. So, we can only get closer from $x < 3$. So, the limit for our function at $x = 3$ <u>does not</u> exist. But based on the convention mentioned earlier, we say the limit exists and is equal to: $\lim\limits_{x \to 3^-} \sqrt{3 - x} = 0$.

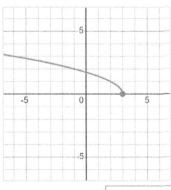

Limit at Infinity

- The limit at infinity describes how a function behaves as the variable approaches positive or negative infinity, meaning it examines how the function approaches a specific value or diverges as x becomes arbitrarily large (both towards negative and positive infinity).

- To evaluate the limit, we analyze the behavior of the function for large values of x.

- It is mathematically denoted as $\lim_{x \to \infty} f(x) = L$, where L is Real Number.

- If x approaches negative infinity, we write it as: $\lim_{x \to -\infty} f(x) = L$.

- If the function grows without bound or oscillates as x approaches infinity, the limit is said to be infinite and undefined, respectively.

Examples:

Example 1. Solve: $\lim_{x \to \infty} \frac{1}{x}$.

Solution: As x becomes larger and larger, the value of $\frac{1}{x}$ approaches 0. Therefore, the limit of $\frac{1}{x}$ as x approaches infinity is 0.

You can see this from the graph of this function:

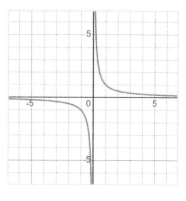

As the we increase the x values, the y or $f(x)$ get closer and closer to 0, no matter if we increase the positive x (getting more positive) or if we decrease the negative x (getting more negative).

Example 2. Solve: $\lim_{x \to \infty} -\frac{5}{x^2} + 5$.

Solution: Similar to previous example, but this time, the graph looks like this:

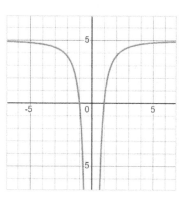

So, as x values increase (or decrease), y values get closer and closer to 5 (Because the $-\frac{5}{x^2}$ part, gets closer to zero as x grows). That's the limit of this function at infinity.

Continuity at a Point

- We say the Function $f(x)$ is continuous at point a, if the limit at a, exists and is equal to the result of $f(x)$ at a: $\lim_{x \to a} f(x) = f(a)$
- If the left-hand limit differs from the right-hand limit, it indicates a discontinuity or some form of jump in the function.
- For a function to be continuous at a point, you should be able to draw its graph without lifting your pen.
- If a function is continuous at <u>all points in its domain</u>, we call it a continuous function. Constant, Linear, Polynomial, Exponential and Trigonometric functions are among the continuous functions.
- There are different types of discontinuity. Here are the common ones:
  - **Point Discontinuity (Removable Discontinuity):** Limit exists, but function undefined or differs from limit at that point.
  - **Jump Discontinuity:** Left-hand and right-hand limits exist but are not equal.
  - **Infinite Discontinuity:** Function approaches infinity near the point.
  - **Essential (oscillating) Discontinuity:** Function oscillates, not settling to a single value or infinity.
- Here we have an example of discontinuous function (jump). As we approach a from left and right, we get two different results. So $f(x)$ can't be equal to $f(a)$.

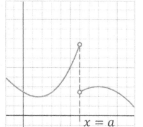

Examples:
Example 1. $\lim_{x \to 2} 7x - 5$

Solution: assuming the function is continuous, it should have a limit equal to $f(2) = 7 \times 2 - 5 = 9$

1.9	1.99	1.999	2	2.001	2.01	2.1
8.3	8.93	8.993	9	9.007	9.07	9.7

Limit indeed approaches 9, so the function is continuous at $x = 2$

Example 2. Show that this function is not continuous for any a values:
$$f(x) = \begin{cases} (2ax + 1)^2 & \text{for } x > 1 \\ x^3 - 2x^2 & \text{for } x \leq 1 \end{cases}$$

Solution: assume each individual piece of this function is continuous, let's see what happens at $x = 1$ itself:

For 1^+: $(2ax + 1)^2 = (2a + 1)^2$, For 1^-: $x^3 - 2x^2 = 1^3 - 2(1)^2 = 1 - 2 = -1$

For $f(x)$ to be continuous, $(2a + 1)^2$ must be equal to -1 which can't be, because square of a number is always positive.

Continuity over an Interval

- Continuity at an interval means that there are no abrupt jumps, breaks, or holes in the function within that interval, and the function can be drawn without lifting the pen or experiencing any discontinuity.

- A function is said to be continuous over an interval if it satisfies three conditions:
  1. The function is defined at every point within the interval.
  2. The limit of the function exists at every point within the interval.
  3. The value of the function at each point within the interval is equal to the limit at that point.

<u>Or</u> mathematically, for function $f(x)$ defined on an interval $[a,b]$. The function is continuous on the interval if:

1. $f(x)$ is defined for all x in $[a,b]$.
2. $\lim\limits_{x \to c} f(x)$ exists for all c in $[a,b]$.
3. For every c in $[a,b]$, $f(c) = \lim\limits_{x \to c} f(x)$.

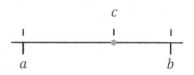

Example:

Example 1. Which of these functions are continuous over $[a,b]$ interval?

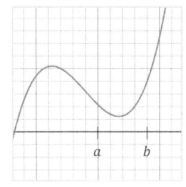 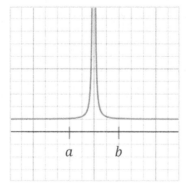 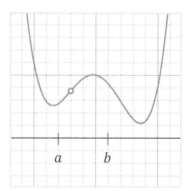

Solution: 2nd and 3rd graph have gaps in them, meaning there are points in $[a,b]$ interval in which $f(x)$ is not equal to f of those points. So, 1st graph is continuous.

Removing Discontinuity

- For functions that are discontinuous at a point, meaning they have continuity but a gap that's making them discontinuous, we can redefine the function into a piecewise function and choose the value for the gap in a way that the new piecewise function becomes continuous.

Examples:

Example 1. Change this function in a way that it's continuous at $x = 1$:
$$f(x) = \frac{x^2 - 1}{x - 1}$$
Solution: For this function, we can't discuss a limit at first because it is undefined, but for every $x \neq 1$, we can simplify the function using factorization to $f(x) = x + 1$, so the only problem is at $x = 1$ itself, which from the $f(x) = x + 1$, would be equals 2 to make the function continuous. So, we reshape our functions into this new piecewise function to come up with a continuous function:
$$\begin{cases} \frac{x^2 - 1}{x - 1} & for \quad x \neq 1 \\ 2 & for \quad x = 1 \end{cases}$$
and now, based on the main condition for a continuous function ($\lim_{x \to a} f(x) = f(a)$), our new function is now continuous, and we removed the discontinuity.

Example 2. Change $f(x) = \frac{x^2-9}{x-3}$ in a way that it's continuous at $x = 3$.

Solution: This function has a discontinuity at $x = 3$ because the denominator becomes *zero*, resulting in an undefined value. To remove the discontinuity, we can factorize the numerator and simplify the function: $f(x) = \frac{(x-3)(x+3)}{x-3}$. Now, the function $f(x) = x + 3$ is a continuous function that fills the gap left by the original function at $x = 3$. The discontinuity has been removed, and the function is defined and continuous at $x = 3$:

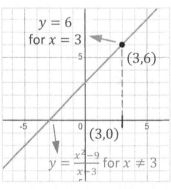

$$g(x) = \begin{cases} \frac{x^2 - 9}{x - 3} & for \quad x \neq 3 \\ 6 & for \quad x = 3 \end{cases}$$

Direct Substitution

- When a limit is continuous at a point, we can find the limit of the function simply by replacing the variable in the function with the given value and computing the resulting expression, algebraically.

- It's important to note that direct substitution may not always work if the function is undefined or if the resulting expression is indeterminate, such as dividing by 0 or encountering an infinity. In such cases, alternative rationalization methods like factoring or applying algebraic manipulation may be required to evaluate the limit.

Examples:

Example 1. Find the answer for $\lim_{x \to 3} 3x^2 + 6x - 5$.

Solution: The domain for our function is $\mathbb{R}$ and is continuous at this domain, so we can see it as an algebraic expression and calculate the limit by substituting x by 3. So, we have: $(3 \times 3^2) + (6 \times 3) - 5 = 40$, and that's the limit of this function at $x \to 3$.

Example 2. $\lim_{x \to 2} x^2 - 3x + 2$.

Solution: Using direct substitution, we simply replace all instances of x with the limiting value, which is 2. So, plugging in 2 into the expression, we get: $(2^2 - 3(2) + 2) = (4 - 6 + 2) = 0$, and that's our limit for $x \to 2$.

Example 3. $\lim_{x \to 2} \frac{3x-5}{12-4x}$.

Solution: This function is defined and continuous at $\mathbb{R} - \{3\}$, so at $x = 2$, we can use direct substitution: $\lim_{x \to 2} \frac{3x-5}{12-4x} = \frac{3 \times 2 - 5}{12 - 4 \times 2} = \frac{1}{4}$.

Example 4. $\lim_{x \to \pi} \sin(x)$.

Solution: Trigonometric functions are among continuous functions, so by substitution of π, we have: $sin(x) = sin(\pi) = 0$.

EffortlessMath.com

Limit Laws

- The limit laws are a set of rules that allow us to simplify and compute limits in various scenarios.
- Limits do not have to be continuous for limit laws to apply, they apply to both continuous and discontinuous functions.
- Here is a list of limit laws:

  a. **Constant Law**: If for every x, $f(x) = b$, then: $\lim\limits_{x \to a} b = b$.

  b. **Identity Law**: Identity law states that: $\lim\limits_{x \to a} x = a$.

  c. **Operational Identities**: Assume $\lim\limits_{x \to a} f(x) = L$ and $\lim\limits_{x \to a} g(x) = M$. and C is a constant. Then, we have the following laws:

  $\lim\limits_{x \to a}[f(x) + g(x)] = \lim\limits_{x \to a} f(x) + \lim\limits_{x \to a} g(x) = L + M$ $\Rightarrow$ (Addition Law)

  $\lim\limits_{x \to a}[f(x) - g(x)] = \lim\limits_{x \to a} f(x) - \lim\limits_{x \to a} g(x) = L - M$ $\Rightarrow$ (Subtraction Law)

  $\lim\limits_{x \to a}[f(x) \times g(x)] = \lim\limits_{x \to a} f(x) \times \lim\limits_{x \to a} g(x) = L \times M$ $\Rightarrow$ (Product Law)

  $\lim\limits_{x \to a} C \times f(x) = C \times \lim\limits_{x \to a} f(x) = C \times L$ $\Rightarrow$ (Coefficient Law)

  $\lim\limits_{x \to a} \dfrac{f(x)}{g(x)} = \dfrac{\lim\limits_{x \to a} f(x)}{\lim\limits_{x \to a} g(x)} = \dfrac{L}{M}$ (For $M \neq 0$) $\Rightarrow$ (Quotient Law)

  $\lim\limits_{x \to a} \dfrac{1}{f(x)} = \dfrac{1}{\lim\limits_{x \to a} f(x)} = \dfrac{1}{L}$ (For $L \neq 0$)

  d. **Power Law**: For $n \in \mathbb{N}$ and $a, b \in \mathbb{R}$:

  $\lim\limits_{x \to a} f(x)^n = \left(\lim\limits_{x \to a} f(x)\right)^n$

  $\lim\limits_{x \to a} x^n = a^n$

  $\lim\limits_{x \to a} b^x = b^a$

  And for $b > 0$, $b \in \mathbb{R}$ and $p(x)$ being a polynomial function, we have:

  $\lim\limits_{x \to a} b^{p(x)} = b^{p(a)}$.

  e. **Root Law**: $\lim\limits_{x \to a} \sqrt[n]{f(x)} = \sqrt[n]{\lim\limits_{x \to a} f(x)}$ (for $n \in \mathbb{N}$)

  for even n values $\{2, 4, 6, \cdots\}$, $\lim\limits_{x \to a} f(x)$ has to be positive.

Example: Using limit laws, simplify the limit: $\lim\limits_{x \to a}(4x^3 + 5x)$.

Solution: using addition law, the limit can be simplified as: $\lim\limits_{x \to a}(4x^3 + 5x) = \lim\limits_{x \to a} 4x^3 + \lim\limits_{x \to a} 5x$, then using coefficient law: $\lim\limits_{x \to a} 4x^3 + \lim\limits_{x \to a} 5x = 4\lim\limits_{x \to a} x^3 + 5\lim\limits_{x \to a} x$

Then using power law: $4\left(\lim\limits_{x \to a} x\right)^3 + 5\lim\limits_{x \to a} x$

Limit Laws Combinations

- Limit laws can be combined to simplify the evaluation of limits. By applying multiple limit laws, we can often simplify complex expressions and determine the limit of a function more easily.

- The idea is to break the expression into smaller parts, evaluate the limits of each part using the appropriate limit laws, and then combine the limits using other limit laws to simplify the overall expression.

Examples:

Example 1. $\lim\limits_{x \to 3} \dfrac{2x-5}{x^2-2x}$.

Solution: Using quotient law, we can spread the limit over both numerator and denominator, so we end up with: $\dfrac{\lim\limits_{x \to 3} 2x-5}{\lim\limits_{x \to 3} x^2-2x}$, which can be calculated by direct substitution: $\dfrac{2 \times 3 - 5}{3^2 - 2 \times 3} = \dfrac{1}{3}$.

Example 2. $\lim\limits_{x \to 1}\left(\sqrt[5]{x^4 - 2x} + (x^2 + x)^4\right)$.

Solution: From addition law, we can: $\lim\limits_{x \to 1} \sqrt[5]{x^4 - 2x} + \lim\limits_{x \to 1}(x^2 + x)^4$.

Now using root and power law: $\sqrt[5]{\lim\limits_{x \to 1} x^4 - 2x} + \left(\lim\limits_{x \to 1}(x^2 + x)\right)^4$.

Which then can be solved as:

$$\sqrt[5]{1^4 - 2 \times 1} + (1^2 + 1)^4 = \sqrt[5]{-1} + 16 = -1 + 16 = 15.$$

Example 3. $\lim\limits_{x \to 1} \dfrac{-x^2+2x+1}{-x+2}$.

Solution: If we look at the graphs of each expression individually, and find their limits at $x = 1$, we find that $-x^2 + 2x + 1$ has a limit of 2 and limit of $-x + 2$ is 1. So, their combination must have a limit of $\dfrac{2}{1} = 2$.

Here you can see the combined graph $\left(\dfrac{-x^2+2x+1}{-x+2}\right)$:

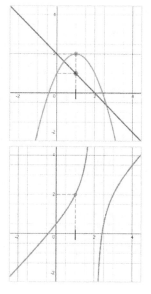

If we approach 1 from left and right, we are going to have a limit of 2.

The Squeeze Theorem

- If a function is trapped (or sandwiched) between two other functions that are getting closer and closer to the same value, then the trapped function also gets closer to that value.

- In the squeeze theorem, the bounding functions must approach the same value as the function being squeezed.

- In mathematical terms, for functions $f(x)$, $g(x)$ and $h(x)$, where the relation $g(x) \leq f(x) \leq h(x)$ is true for neighborhood of point a (not necessarily at a itself), we define the squeeze theorem as:

If $\lim\limits_{x \to a} g(x) = \lim\limits_{x \to a} h(x) = L$, then: $\lim\limits_{x \to a} f(x) = L$.

Example:

Example 1. Find the answer to this limit: $\lim\limits_{x \to 0} x \sin\left(\frac{1}{x}\right)$.

Solution: We know from unit circle that $-1 \leq \sin\left(\frac{1}{x}\right) \leq 1$.

We multiply the inequation by x to produce the function in the problem, but x can be negative too, so we have two possibilities:

For $x > 0$: $-x \leq x \sin\left(\frac{1}{x}\right) \leq x$.

For $x < 0$: $-x \geq x \sin\left(\frac{1}{x}\right) \geq x$.

As you can see from the graph, in both cases, $x \sin\left(\frac{1}{x}\right)$ is between x and $-x$ graphs, and the limit for x and $-x$ for x approaching 0 is 0. So according to the squeeze theorem, we can conclude that $\lim\limits_{x \to 0} x \sin\left(\frac{1}{x}\right)$ is also going to be 0.

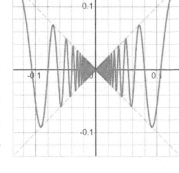

Example 2. Calculate $\lim\limits_{x \to 0} x \frac{1}{\lfloor x \rfloor}$.

Solution: Based on floor functions properties, we have: $\frac{1}{x} - 1 < \lfloor \frac{1}{x} \rfloor \leq \frac{1}{x}$, if we multiply this by x, we get the function in the problem, so:

For $x > 0$: $x\left(\frac{1}{x} - 1\right) < x \lfloor \frac{1}{x} \rfloor \leq 1$ and for $x < 0$: $x\left(\frac{1}{x} - 1\right) > x \lfloor \frac{1}{x} \rfloor \geq 1$.

$x\left(\frac{1}{x} - 1\right)$ is equals to $1 - x$, and our function is between $1 - x$ and 1 in both cases. Since $\lim\limits_{x \to 0}(1 - x) = \lim\limits_{x \to 0} 1 = 1$, according to squeeze theorem, the answer for our limit is also 1.

Indeterminate and Undefined

- When dealing with limits, we might encounter several different situations where we have to deal with zero as an absolute number, values really close to zero, or negative and positive infinity.

- For numbers that are close to zero: if it's bigger than *zero* we show them with 0^+ (like 0.0000002), and if its smaller than zero, 0^- (like -0.0000002). Same notation applies for other numbers, for example 1^+ means numbers slightly bigger than one.

- There is a distinction between "undefined" and "indeterminate" in mathematics. "Undefined" refers to a situation where a mathematical expression or operation does not have a valid or meaningful interpretation. It typically occurs when certain mathematical rules or definitions are violated, leading to inconsistencies or contradictions. As an example, zero in denominator is always undefined.

- On the other hand, "indeterminate" refers to a situation where a mathematical expression or limit does not provide enough information to determine a unique value. Meaning it could have a finite and defined answer, an undefined answer or infinity. For example: $\infty - \infty$

- Rationalizing is the process of modifying an irrational expression by multiplying it by a suitable form of 1 to eliminate any radical (square root) or complex denominator to express the quantity in a more simplified or standard form.

Examples:

Example 1. Evaluate the limit: $\lim\limits_{x \to 0^+} \frac{1}{x}$.

Solution: as x gets closer to zero from the positive side, $\frac{1}{x}$ becomes larger and larger, because dividing by a very small positive number yields a very large number. So, the limit is infinity.

Example 2. Evaluate the limit: $\lim\limits_{x \to 0^+} \frac{\sqrt{x+1}-1}{x}$.

Solution: as $x \to 0$, both the numerator and denominator approach zero, the function approaches $\frac{0}{0}$. Let's try to rationalize the expression: Multiply the numerator and denominator by the conjugate of the numerator to eliminate the square root: $\frac{\sqrt{x+1}-1}{x} \times \frac{\sqrt{x+1}+1}{x\sqrt{x+1}+1} = \frac{(x+1)-1}{x(x\sqrt{x+1}+1)} = \frac{x}{x(x\sqrt{x+1}+1)} = \frac{1}{x\sqrt{x+1}+1}$, which approaches $\frac{1}{2}$.

Infinity Cases

- Here we have a set of different cases for when numbers, small numbers near 0, 0 itself and negative/positive infinity meet up. Although if you use logic, you should be able to generate the answers to most of these cases by yourself, if not all.

  - $L > 0, \in \mathbb{R}: \frac{L}{0^+} = +\infty,\ \frac{L}{0^-} = -\infty,\ L \times (+\infty) = +\infty,\ L \times (-\infty) = -\infty$
  - $L < 0, \in \mathbb{R}: \frac{L}{0^+} = -\infty,\ \frac{L}{0^-} = +\infty,\ L \times (+\infty) = -\infty,\ L \times (-\infty) = +\infty$
  - $L \in \mathbb{R}: +\infty \pm L = +\infty,\ -\infty \pm L = -\infty,\ -\infty - \infty = -\infty,\ +\infty + \infty = +\infty$
  - $(-\infty) \times (+\infty) = -\infty,\ (-\infty) \times (-\infty) = +\infty,\ (+\infty) \times (+\infty) = +\infty$
  - $\frac{+\infty}{0^+} = +\infty,\ \frac{+\infty}{0^-} = -\infty,\ \frac{-\infty}{0^+} = -\infty,\ \frac{-\infty}{0^-} = +\infty$
  - $\begin{cases} |a| > 1 & \Rightarrow\ a^{+\infty} = +\infty \\ -1 < |a| < 1 & \Rightarrow\ a^{-\infty} = +\infty \end{cases}$

- Here are some additional cases that can occur:

  - $\frac{0}{0^\pm} = 0$
  - $\frac{0^\pm}{\infty} = 0$
  - $\frac{x \in \mathbb{R}}{\infty} = 0$
  - $\frac{0^\pm}{x \in \mathbb{R},\ x \neq 0} = 0$
  - $0 \times \infty = 0$
  - $0 \times 0^\pm = 0$
  - $(0^\pm)^0 = 1$

Examples:

Example 1. Evaluate $\lim\limits_{x \to 0^+} \frac{\lfloor x \rfloor}{x}$

Solution: The floor function rounds a number down to the nearest integer, since 0^+ means values slightly greater than 0, $\lfloor 0^+ \rfloor$ equals 0, and we have $\frac{0}{0^+} = 0$

Example 2. Answer $\lim\limits_{x \to 3^+} \frac{x-3}{\lfloor x \rfloor + \lfloor -x \rfloor}$

Solution: $= \frac{3^+ - 3}{\lfloor 3^+ \rfloor + \lfloor -(3^+) \rfloor} = \frac{0^+}{3 - 4} = \frac{0^+}{-1} = 0$

Example 3. Find the limit: $\lim\limits_{x \to 3^+} \frac{\lfloor x \rfloor - 3}{x^2 - 9}$

Solution: by placing 3^+, we have: $\frac{\lfloor 3^+ \rfloor - 3}{9^+ - 9} = \frac{3 - 3}{0^+} = \frac{0}{0^+} = 0$

Trigonometric Limits

- Similar to limit law $\lim\limits_{x \to a} f(x) = a$ mentioned before, we have this set of laws for trigonometric functions: (x in radians)

  - $\lim\limits_{x \to a} \sin(x) = \sin(a)$
  - $\lim\limits_{x \to a} \tan(x) = \tan(a)$
    $(a \neq k\pi + \frac{\pi}{2}, k \in z)$
  - $\lim\limits_{x \to a} \cos(x) = \cos(a)$
  - $\lim\limits_{x \to a} \cot(x) = \cot(a)$
    $(a \neq k\pi, k \in z)$

- Same as on real number line, for x angles approaching the angle a, we have:

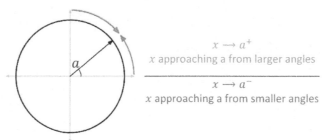

Examples:

Example 1. $\lim\limits_{x \to \frac{\pi}{2}} \frac{2\sin(x)+\cos(x)}{\cos(2x)-3\sin(3x)}$.

Solution: From quotient law, we have $\dfrac{\lim\limits_{x \to \frac{\pi}{2}}(2\sin(x)+\cos(x))}{\lim\limits_{x \to \frac{\pi}{2}}(\cos(2x)-3\sin(3x))}$ and from addition and product law, we have: $\dfrac{2\lim\limits_{x \to \frac{\pi}{2}}\sin(x)+\lim\limits_{x \to \frac{\pi}{2}}\cos(x)}{\lim\limits_{x \to \frac{\pi}{2}}\cos(2x)-3\lim\limits_{x \to \frac{\pi}{2}}\sin(3x)} = \dfrac{2 \times 1 + 0}{-1 - 3(-1)} = 1$.

Example 2. $\lim\limits_{x \to 0^-} \frac{[\sin(x)]}{x}$.

Solution: We know from the unit circle that angles smaller than 0 are placed in 4th region of unit circle, which has negative values for sine. So:

$\lim\limits_{x \to 0^-} \frac{[\sin x]}{x} = \frac{[\sin 0^-]}{0^-} = \frac{[0^-]}{0^-} = \frac{-1}{0^-} = +\infty$. So, the function doesn't have a limit.

Example 3. $\lim\limits_{x \to 2}[\cos(x)]$.

Solution: x is in radians, and each radian is around $57°$, so 2 radians is in the second region of unit circle, with $-1 < \cos(x) < 0$, so the answer using the properties of floor function, is: -1

Rationalizing Trigonometric Functions

- We may encounter a $\frac{0}{0}$ situation with trigonometric functions. There are a set of laws for this type of problem that are derived from geometry and unit circle.
- Let's assume x is an angle in radians. For angles $x \in \left(\frac{-\pi}{2}, \frac{\pi}{2}\right) - \{0\}$, meaning angles in 1st or 4th region of unit circle, we have:
$cos(x) < \frac{sin(x)}{x} < 1$ and using squeeze theorem, we can conclude:
  - $\lim\limits_{x \to 0} \frac{sin(x)}{x} = 1$
  - $\lim\limits_{x \to 0} \frac{x}{sin(x)} = 1$
  - $\lim\limits_{x \to 0} \frac{tan(x)}{x} = 1$
  - $\lim\limits_{x \to 0} \frac{x}{tan(x)} = 1$
- In a more general sense, we can rewrite the laws above as:
  - $\lim\limits_{x \to 0} \frac{sin(ax)}{x} = a$
  - $\lim\limits_{x \to 0} \frac{x}{sin(ax)} = \frac{1}{a}$
  - $\lim\limits_{x \to 0} \frac{tan(ax)}{x} = a$
  - $\lim\limits_{x \to 0} \frac{x}{tan(ax)} = \frac{1}{a}$

Examples:

Example 1. $\lim\limits_{x \to 0} \frac{sin^2(3x)}{x^2}$.

Solution: $\lim\limits_{x \to 0} \frac{sin^2(3x)}{x^2} = \lim\limits_{x \to 0} \left(\frac{sin(3x)}{x}\right)^2 = 3^2 = 9$

Example 2. $\lim\limits_{x \to 0} \frac{\sqrt{cos(3x)} - \sqrt{cos(x)}}{x^2}$.

Solution: Using a proper 1 to simplify the radical with, and the identity $cos(a) - cos(b) = -2 sin\left(\frac{a+b}{2}\right) sin\left(\frac{a-b}{2}\right)$:

$\lim\limits_{x \to 0} \frac{\sqrt{cos(3x)} - \sqrt{cos(x)}}{x^2} \times \frac{\sqrt{cos(3x)} + \sqrt{cos(x)}}{\sqrt{cos(3x)} + \sqrt{cos(x)}} = \lim\limits_{x \to 0} \frac{cos(3x) - cos(x)}{x^2(\sqrt{cos(3x)} + \sqrt{cos(x)})} = \lim\limits_{x \to 0} \frac{-2 sin\left(\frac{3x-x}{2}\right) sin\left(\frac{3x+x}{2}\right)}{x^2(\sqrt{cos(3x)} + \sqrt{cos(x)})}$

$= -2 \lim\limits_{x \to 0} \left(\frac{sin(x)}{x}\right) \left(\frac{sin(2x)}{x}\right) \times \left(\frac{1}{\sqrt{cos(3x)} + \sqrt{cos(x)}}\right) = -2 \times 1 \times 2 \times \frac{1}{2} = -2$

Example 3. $\lim\limits_{x \to 0} \frac{x - sin(2x)}{x + sin(3x)}$.

Solution: Simplify by $\frac{x}{x} = 1$: $\lim\limits_{x \to 0} \frac{\frac{x - sin(2x)}{x}}{\frac{x + sin(3x)}{x}} = \lim\limits_{x \to 0} \frac{\frac{x}{x} - \frac{sin(2x)}{x}}{\frac{x}{x} + \frac{sin(3x)}{x}} = \frac{1 - 2}{1 + 3} = -\frac{1}{4}$.

Example 4. $\lim\limits_{x \to 0^-} \frac{sin(x) - sin(5x)}{\sqrt{1 - cos(4x)}}$.

Solution: Using the identity $sin(a) - sin(b) = 2 cos\left(\frac{a+b}{2}\right) sin\left(\frac{a-b}{2}\right)$:

$\lim\limits_{x \to 0} \frac{2 sin\left(\frac{x-5x}{2}\right) cos\left(\frac{x+5x}{2}\right)}{\sqrt{2 sin^2(2x)}} = \lim\limits_{x \to 0^-} \frac{-2 sin(2x) cos(3x)}{|sin(2x)|\sqrt{2}} = \lim\limits_{x \to 0^-} \frac{-2 sin(2x) cos(3x)}{-sin(2x)\sqrt{2}} =$

$\lim\limits_{x \to 0^-} \frac{2 cos(3x)}{\sqrt{2}} = \frac{2 \times 1}{\sqrt{2}} = \sqrt{2}$.

Algebraic Manipulation

- Algebraic manipulation involves performing operations such as distribution and combining like terms to simplify or transform algebraic expressions. It aims to rearrange the terms and factors in an expression without changing the overall value or meaning.

- Indeterminate limits like the ones with numerator and denominator very close to zero $\left(\frac{0^{\pm}}{0^{\pm}}\right)$ are considered vague and we have to manipulate the expressions to get rid of the piece that's causing the indeterminate.

- If we are dealing with $\lfloor f(x) \rfloor$ or $|f(x)|$, which are floor function and absolute value function, and we encounter an indeterminate scenario, we can find the sign of the expressions inside the bracket, and based on that, get rid of brackets.

- For vague radical functions, we can multiply both numerator and denominator at something that can neutralize the radical. (basically, multiplying by 1)

Examples:

Example 1. $\lim\limits_{x \to 2} \frac{x^2 - 4}{x - 2}$.

Solution: If we place 2 for x, we get an indeterminate $\left(\frac{0}{0}\right)$, but for $x \to 2$, we don't actually reach 2 itself, so we can simplify the algebraic expression: $\frac{x^2-4}{x-2} = \frac{(x-2)(x+2)}{x-2} = x+2$, and for $x \to 2$ the answer for limit is 4.

Example 2. $\lim\limits_{x \to 1} \frac{4x^3 + 3x^2 - 7}{x^3 - 1}$.

Solution: We reach another $\left(\frac{0}{0}\right)$ when placing 1 in the expression. If we factor the expression to isolate $(x - 1)$, which is causing the indeterminate situation, we have: $\lim\limits_{x \to 1} \frac{(x-1)(4x^2+7x+7)}{(x-1)(x^2+x+1)} = \frac{18}{3} = 6$.

Example 3. $\lim\limits_{x \to 0^-} \frac{x - |x|}{\lfloor x+1 \rfloor - x}$.

Solution: For x smaller than *zero*, $\lfloor x+1 \rfloor = 0$ and $|x| = -x$, so we have $\frac{x-|x|}{\lfloor x+1 \rfloor - x} = \frac{x-(-x)}{0-x} = \frac{2x}{-x} = -2$.

Example 4. $\lim\limits_{x \to 1} \frac{x^2 - 1}{\sqrt{x} - 1}$.

Solution: $\lim\limits_{x \to 1} \frac{x^2-1}{\sqrt{x}-1} \times \frac{\sqrt{x}+1}{\sqrt{x}+1} = \lim\limits_{x \to 1} \frac{(x-1)(x+1)\sqrt{x}+1}{x-1} = \lim\limits_{x \to 1}(x+1)(\sqrt{x}+1) = 4$.

Redefining function's value

- Another approach to solve vague limits, is to change the x in $f(x)$ into a new value that makes the previously undefined limit solvable.

- So, in mathematical terms, if we use $(t + a)$ instead of x, we going to have: $\lim_{x \to a} f(x) = \lim_{t \to 0} f(t + a)$.

- Notice that instead of $x \to a$, we now have $t \to 0$, meaning by changing the x, we are also changing the value that the limit approaches. This is because if $x = t + a$, if x gets closer to a (from original limit), t must approach 0.

Examples:

Example 1. $\lim_{x \to 1} \frac{\sin(\pi x)}{x^2 - 1}$.

Solution: $x = t + 1$ and from trigonometry, we know that $\sin(\pi + \theta) = -\sin(\theta)$, so: $\lim_{x \to 1} \frac{\sin(\pi x)}{x^2 - 1} = \lim_{x \to 1} \frac{\sin(\pi x)}{(x-1)(x+1)} = \lim_{t \to 0} \frac{\sin(\pi(1+t))}{t(t+2)} = \lim_{t \to 0} \frac{\sin(\pi + \pi t)}{t(t+2)}$

$= \lim_{t \to 0} \frac{-\sin(\pi t)}{t(t+2)} = \lim_{t \to 0} \left(\frac{\sin(\pi t)}{t}\right) \times \left(\frac{-1}{t+2}\right) = \pi \times \left(-\frac{1}{2}\right) = -\frac{\pi}{2}$

Example 2. $\lim_{x \to 1} \frac{2\sqrt{x} - 3\sqrt[3]{x} + 1}{(1 - \sqrt{x})(1 - \sqrt[3]{x})}$.

Solution: This time, it's better to assign t to: $x = t^6$ so both radical indexes can be interactable by t^6 and therefore as $x \to 1$, $t \to 1$ as well. Now to solve the limit:

$\lim_{t \to 1} \frac{2\sqrt{t^6} - 3\sqrt[3]{t^6} + 1}{(1 - \sqrt{t^6})(1 - \sqrt[3]{t^6})} = \lim_{t \to 1} \frac{2t^3 - 3t^2 + 1}{(1 - t^3)(1 - t^2)}$

$= \lim_{t \to 1} \frac{(t-1)^2 (2t+1)}{(1-t)(1+t+t^2)(1-t)(1+t)} = \lim_{t \to 1} \frac{2t+1}{(1+t+t^2)(1+t)} = \frac{3}{6} = \frac{1}{2}$

Example 3. $\lim_{x \to a} \frac{\sin(x-a)}{x^2 - a^2}$.

Solution: Let's say $t = x - a$, so the new limit is: $\lim_{t \to 0} \frac{\sin(t)}{t(t+2a)}$.

We know that $\lim_{t \to 0} \frac{\sin(t)}{t} = 1$, so we rewrite our limit as:

$\lim_{t \to 0} \frac{\sin(t)}{t(t+2a)} = \lim_{t \to 0} \frac{\sin(t)}{t} \times \lim_{t \to 0} \frac{1}{(t+2a)} = 1 \times \frac{1}{2a} = \frac{1}{2a}$.

Rationalizing Infinite Limits

- Sometimes we encounter an ambiguous situation (indeterminate) that is not of $\frac{0}{0}$ type but includes ∞ instead (like: $0 \times \infty$, $\frac{\infty}{\infty}$, $\cdots$). We need to somehow change it to $\frac{0}{0}$ type in order to be able to solve it, then we can use the tricks we previously explored to find the limit.

- General algebraic principles regarding infinity are as follows:

    1) Let's assume $\lim\limits_{x \to a} f(x) = +\infty$ and $\lim\limits_{x \to a} g(x) = +\infty$, then we have:

    $$\lim_{x \to a}(f(x) \times g(x)) = +\infty \text{ and } \lim_{x \to a}(f(x) + g(x)) = +\infty.$$

    In the first principle, if $L = 0$, the $\lim\limits_{x \to a} f(x) \times g(x)$, we get ambiguity of $0 \times \infty$ type. We can no longer solve this case by number 1 coefficient principle, what we need to do is to send the function that's causing infinity to denominators, like so: $\lim\limits_{x \to a} f(x) \times g(x) = \lim\limits_{x \to a} \frac{g(x)}{\frac{1}{f(x)}}$, so as we know, from $\frac{real\ number}{\infty} = 0$, we get a $\frac{0}{0}$ situation and we can solve the limit.

    Similarly, if we get $\frac{\infty}{\infty}$, we can inverse both numerator and denominator to reach $\frac{0}{0}$ and be able to solve it.

    2) Sometimes all we need to do is simplify the expressions and get out of infinity ambiguity.

Examples:

Example 1. $\lim\limits_{x \to 0} \sin(3x) \times \cot(2x)$.

Solution: We get ambiguous $0 \times \infty$, with $\cot(2x)$ causing infinity, so as mentioned above: $\lim\limits_{x \to 0} \sin(3x) \times \cot(2x) = \lim\limits_{x \to 0} \frac{\sin(3x)}{\frac{1}{\cot(2x)}} = \lim\limits_{x \to 0} \frac{\sin(3x)}{\tan(2x)} = \lim\limits_{x \to 0} \frac{\frac{\sin(3x)}{x}}{\frac{\tan(2x)}{x}} = \frac{3}{2}$.

Example 2. $\lim\limits_{x \to 2}\left(\frac{1}{x-2} - \frac{4}{x^2-4}\right)$.

Solution: We can simplify the fraction, which is indeterminate $\infty - \infty$:

$\lim\limits_{x \to 2}\left(\frac{1}{x-2} - \frac{4}{x^2-4}\right) = \lim\limits_{x \to 2} \frac{(x+2)-4}{x^2-4} = \lim\limits_{x \to 2} \frac{x-2}{(x-2)(x+2)} = \lim\limits_{x \to 2} \frac{1}{x+2} = \frac{1}{4}$.

Chapter 4: Practices

✎ **Find the given limits from the graph of $f(x)$:**

1) $\lim\limits_{x \to 5} f(x) =$

2) $\lim\limits_{x \to 4} f(x) =$

3) $\lim\limits_{x \to 1^+} f(x) =$

4) $\lim\limits_{x \to 0} f(x) =$

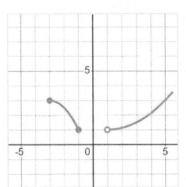

✎ **Find the limit of $g(x)$ for $x \to 2$ if it exists, using the table.**

5) $g(x) = \dfrac{3x+1}{x-2};\ \lim\limits_{x \to 2} g(x) =$

x	1.9	1.99	2	2.01	2.1
$g(x)$					

✎ **Which function is continuous at the interval $[3, 6)$?**

6) $f(x) = x^2 + 3x - 2$, or $g(x) = \dfrac{1}{x-4}$, or $h(x) = \sqrt{x-5}$

✎ **Solve.**

7) $\lim\limits_{x \to -2} 2^{(x^2+3x)} =$

8) $\lim\limits_{x \to 1} \dfrac{x^2-1}{x^3-1} =$

9) $\lim\limits_{x \to 0} \dfrac{\sin(2x)}{\sqrt{1+\tan(x)} - \sqrt{1-\tan(x)}} =$

10) $\lim\limits_{x \to \infty} \dfrac{50}{3x^2} =$

Effortless Math Education

11) $f(0)$ in a way that the function $f(x) = \frac{\sin x}{x - |x|}$ is continuous at point $x = 0$.

✎ Solve the following limits.

12) $\lim\limits_{x \to 2} \frac{x^3 - 6x^2 + 8x}{x^2 - 3x + 2} =$

13) $\lim\limits_{x \to 0} \frac{\sqrt{x^4 + 3x^2} - 2\sqrt{x}}{\sqrt{x}} =$

✎ Assuming for $-1 \leq x \leq 1$, we have $\sqrt{3 - 2x^2} \leq f(x) \leq \sqrt{3 + x^2}$.

14) Find $\lim\limits_{x \to 0} \frac{1}{f(x)} =$

✎ Evaluate the limits:

15) $\lim\limits_{x \to 2^-} \frac{|x^2| - 4}{2 - x} =$

16) $\lim\limits_{x \to 1^-} \frac{\lfloor x \rfloor}{x - 1} =$

17) $\lim\limits_{x \to 0} \frac{\cos^2 x - \sqrt{\cos x}}{x^2} =$

18) $\lim\limits_{x \to 1} x - \lfloor x \rfloor =$

✎ Find the midpoint of this interval:

19) $\left(\frac{90}{7}, \frac{46}{3} \right) =$

✎ Evaluate this limit:

20) $\lim\limits_{x \to 3^+} \lfloor 2x - 6 \rfloor - \lfloor 4x + 7 \rfloor + \lfloor x \rfloor =$

Chapter 4: Answers

1) 3

2) 2

3) 1

4) Doesn't exist.

5) Doesn't exist.

6) $f(x)$

7) $\frac{1}{4}$

8) $\frac{2}{3}$

9) 2

10) 0

11) $\begin{cases} \frac{\sin(x)}{x-|x|} & \text{for } x < 0 \\ \frac{1}{2} & \text{for } x = 0 \end{cases}$

12) -4

13) -2

14) $\frac{\sqrt{3}}{3}$

15) No limit

16) 0

17) $-\frac{3}{4}$

18) No limit

19) $\frac{592}{42} = 14.09$

20) -15

CHAPTER 5
Derivative

Math topics that you'll learn in this chapter:

- ☑ Derivative Introduction
- ☑ Average and Instantaneous Rates of Change
- ☑ Derivative of a Function
- ☑ Rules of differentiation
- ☑ Derivative of Trigonometric Functions
- ☑ Power Rule
- ☑ Product rule
- ☑ Quotient Rule
- ☑ Chain Rule
- ☑ Derivative of Radicals
- ☑ Derivative of Logarithms
- ☑ L'Hôpital's rule
- ☑ Differentiability
- ☑ Second Derivative and Minimum/Maximum
- ☑ Curve Sketching Using Derivatives
- ☑ Differentiating Inverse Functions
- ☑ Optimization Problems
- ☑ Implicit Differentiation
- ☑ Related Rates
- ☑ Derivative of Inverse Function

Derivative Introduction

- Derivatives are a fundamental part of calculus, a branch of mathematics that deals with rates of change and the relationship between a function and its slope (tangent line).

- To understand derivatives, we need to utilize the concept of the limit.

- Their applications are vast and can be found in various fields, such as physics, engineering, economics, and more.

- They can be calculated for various functions, including polynomial, exponential, trigonometric, and logarithmic functions.

- There are various notations for derivatives in different branches of science, for example: $Df(x), \frac{d}{dx}(y), f'(x), \cdots$.

- "Differentiation" and "taking the derivative" essentially refer to the same mathematical process.

- Derivatives are essential in measuring change. For example, they are used to calculate instantaneous rates of change, which are crucial in determining rate of population growth or reactions in chemical processes.

- In physics, derivatives are used to analyze motion-related phenomena. They provide crucial information about the dynamics of objects in motion, such as velocity and acceleration. The derivative of displacement gives velocity, and similarly, the derivative of velocity yields acceleration. So, by taking the derivative of the car's position with respect to time, we can calculate its velocity, which tells us how fast the car is moving at a specific instant. Similarly, by taking the derivative of velocity with respect to time, we can determine the car's acceleration, which indicates how quickly the car's velocity is changing.

- One of the main applications of derivatives is in optimization problems. By finding the derivative of a function and setting it equal to zero, one can locate the critical points or extrema of a function. This information is used in maximizing or minimizing a function, ensuring efficiency and effectiveness in many real-world scenarios. For instance, finding the shortest path for transportation networks.

- Derivatives can be used to calculate risk, analyze physiological variables, model disease spread rates, and study environmental changes.

Average and Instantaneous Rates of Change

- The concept of derivatives is used to measure how a function changes at any given point. Two important measures of this change are the average rate of change and the instantaneous rate of change.

- The average rate of change calculates the average rate at which a function changes over a specific interval. It is determined by "dividing the change in the function's output values by the difference in the corresponding input values".

- On the other hand, the instantaneous rate of change refers to the rate of change at a specific point on the function. It is determined by taking the derivative of the function at that point. The derivative measures the slope of the tangent line to the curve at a particular point, indicating how fast the function is changing at that specific moment.

- Both the average and instantaneous rates of change provide crucial insights into the behavior of a function. The average rate shows the overall trend of change over a specific interval, while the instantaneous rate highlights the specific rate of change at any given point.

Examples:

Example 1. Let's say you're driving from point A to point B, which are 100 miles apart. You start at 9:00 AM and arrive at 11:00 AM. Find the average rate of change.

Solution: To find the average rate of change, divide the change in the function's output values, by the difference in the corresponding input values. The average rate of change of your distance traveled during this period would be $\left(\frac{100\ mi}{2h}\right) = 50$ miles per hour.

Example 2. In the previous example, how can you find the instantaneous speed of car at any moment?

Solution: In the previous example, one does not simply drive their car at a fixed speed, and it is constantly changing by environmental conditions. But at any given time (seconds, minutes, ⋯) the car has a specific instantaneous speed. The speedometer of the car shows this using mechanical means.

bit.ly/47E3D55
Find more at

The Derivative of a Function

- The derivative can also be interpreted as the instantaneous rate of change of the function with respect to its independent variable.
- The derivative is computed by finding the limit of the average rate of change of the function as the interval over which it is calculated becomes really-really small (infinitesimally small).

Examples:

Example 1. Let's consider the function $f(x) = 2x^2$.

Solution: To find the derivative of the function $f(x) = 2x^2$ using the limit of the average rate of change, we can follow these steps:

1) Calculate the average rate of change for $f(x)$ over a small interval, h. The average rate of change for a function $f(x)$ over an interval h is given by: $\frac{(f(x+h)-f(x))}{(x+h)-x} = \frac{(f(x+h)-f(x))}{h}$. As h approaches 0, this average rate becomes the instantaneous rate of change or the derivative.

   For the function $f(x) = 2x^2$, let's calculate the average rate of change over an interval h as follows: $\frac{(f(x+h)-f(x))}{h} = \frac{(2(x+h)^2 - 2x^2)}{h} = \frac{(2(x^2+2xh+h^2)-2x^2)}{h} = \frac{2x^2+4xh+2h^2-2x^2}{h} = \frac{4xh+2h^2}{h} = 4x + 2h$

2) Take the limit as h approaches 0: $\lim_{h \to 0}(4x + 2h) = 4x + 0 = 4x$.

   Therefore, the derivative of the function $f(x) = 2x^2$ using the limit of the average rate of change is $4x$.

Example 2. Find the derivative of $sin(x)$.

Solution: We use the definition of the derivative and apply trigonometric identities to simplify the expression. Start with $\frac{sin(x+h)-sin(x)}{h}$. using the trigonometric identity $sin(x+h) = sin(x)cos(h) + cos(x)sin(h)$, we have: $\frac{sin(x)cos(h)+cos(x)sin(h)-sin(x)}{h}$. Split into two terms: $\frac{sin(x)cos(h)-sin(x)}{h} + \frac{cos(x)sin(h)}{h} = \frac{sin(x)cos(h-1)}{h} + \frac{cos(x)sin(h)}{h}$. Now, we take this limit: $\lim_{h \to 0} \frac{sin(x)cos(h-1)}{h} + \frac{cos(x)sin(h)}{h}$, and since $sin(x)$ and $cos(x)$ are not variables of h, so we can move them out of the limit:

$\left(sin(x) \times \lim_{h \to 0} \frac{cos(h-1)}{h}\right) + \left(cos(x) \times \lim_{h \to 0} \frac{sin(h)}{h}\right) = (0) + (cos(x) \times 1) = cos(x)$

Derivative of Trigonometric Functions

- For trigonometric functions like $sin(x)$, there are specific derivative rules that are used to determine how the function is changing with respect to the independent variable (in this case, x)

- Here are the results of same procedure on main trigonometric functions:

  - $(sin(x))' = cos(x)$
  - $(tan(x))' = 1 + tan^2(x)$
  - $(cos(x))' = -sin(x)$
  - $(cot(x))' = -(1 + cot^2(x))$

- In case we have $f(x)$ as variable of a trigonometric function instead of just x, we need to multiply $f'(x)$ to our answer: $(sin(f(x)))' = cos(x) \times f'(x)$.

  This is because of the "chain rule" mentioned earlier.

Examples:

Example 1. Find the derivative of $f(x) = cos(2x)$.

Solution: We need to use the chain rule, because we have $2x$ in front of cosine, not x. So, we have: $(2x)' = 2$ and the $cos(2x)$ has a derivative of: $-sin(2x)$, so the answer is: $-2 sin(2x)$.

Example 2. Find the derivative of $g(x) = 1 + cot(x)$.

Solution: The derivative of a constant is *zero*, so we only need to find the derivative of $cot(x)$, which is: $-(1 + cot^2(x))$, which if simplified, is actually equal to: $-csc^2(x)$.

Example 3. Find the derivative of $h(x) = 3 sin(x) + cos(2x)$.

Solution: Derivative of $3 sin(x) = 3 cos(x)$.
$(cos(2x))' = -2 sin(2x)$, so, when combined, we have:
$3 cos(x) - 2 sin(2x)$.

Example 4. Find the derivative of $i(x) = sin(cos(3x))$.

Solution: starting from the outer function and treating what's in each using chain rule, we have: $(sin(cos(3x)))' = cos(cos(3x)) \times (cos(3x))' = (cos(cos(3x))) \times (-sin(3x)) \times 3 = -3(sin(3x))(cos(cos(3x)))$

Power Rule

- The Power Rule of differentiation is a fundamental rule that allows us to find the derivative of a power function. If $f(x) = x^n$, where n is a constant, the Power Rule states that the derivative of $f(x)$ with respect to x is:

$$f'(x) = nx^{n-1}$$

In simpler terms, multiply the original exponent (n) by the variable (x) and for the new exponent of x, place ($n - 1$), even if it gives negative results or zero.

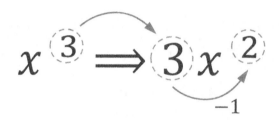

Examples:

Example 1. Consider $f(x) = x^3$.

Solution: Applying the power rule, we differentiate the function:

$f'(x) = 3x^{3-1} = 3x^2$.

Example 2. Consider $g(x) = 5x^{-2}$.

Solution: $g'(x) = 5(-2x^{-2-1}) = -10x^{-3} = -\frac{10}{x^3}$.

Example 3. Consider $h(x) = 2x^4$.

Solution: $h'(x) = 2(4x^{4-1}) = 8x^3$.

Example 4. $f(x) = 3x^4 - 2x^3 + 7x^2 - 5x + 9$.

Solution: $3(4x^{4-1}) - 2(3x^{3-1}) + 7(2x^{2-1}) - 5 + 0 = 12x^3 - 6x^2 + 14x - 5$.

Example 5. $f(x) = -3x^4 + 5x^{-2} - 7x^2$.

Solution: $f'(x) = -12x^3 - 10x^{-3} - 14x = -\left(12x^3 + \frac{10}{x^3} + 14x\right)$.

Example 6. $f(x) = \sqrt{x^3}$.

Solution: Rewrite $f(x)$ to $x^{\frac{3}{2}}$, then: $f'(x) = \frac{3}{2}x^{\frac{3}{2}-1} = \frac{3}{2}x^{\frac{1}{2}} = \frac{3}{2}\sqrt{x}$.

Product Rule

- The Product Rule is a fundamental rule for finding the derivative of the product of two functions. Given two functions, $f(x)$ and $g(x)$, the Product Rule states that the derivative of their product, denoted as $(f(x) \times g(x))'$, is equal to $f'(x) \times g(x) + f(x) \times g'(x)$. In other words, to find the derivative of a product, we take the derivative of the first function times the second function, then add it to the first function times the derivative of the second function.

Examples:

Example 1. Let's find the derivative of the function $f(x) = x^2 \times sin(x)$.

Solution: Using the Product Rule, we have:

$f'(x) = (x^2)'(sin(x)) + (x^2)(sin(x))' = (2x \times sin(x)) + (x^2 \times cos(x))$

Example 2. Consider the function $g(x) = 3x^2(x^2 - 4x)$.

Solution: Using the Product Rule, we have:

$g'(x) = 6x(x^2 - 4x) + 3x^2(2x - 4)$

$= 6x^3 - 24x^2 + 6x^3 - 12x^2$

$= 12x^3 - 36x^2 = 12x^2(x - 3)$

We can also solve the problem from another method, which is simplifying the function first and then taking the derivative, which yields the same result:

$g(x) = 3x^2(x^2 - 4x) = 3x^4 - 12x^3$, so: $(3x^4 - 12x^3)' = 12x^3 - 36x^2 = 12x^2(x - 3)$

Example 3. Suppose we have the function $f(x) = sin(x) cos(x)$.

Solution: $f'(x) = (sin(x) cos(x))' = (sin(x))' cos(x) + sin(x) (cos(x))'$

$= cos(x) cos(x) + sin(x) (-sin(x)) = cos^2(x) - sin^2(x) = cos(2x)$

Or using the trigonometric identity, $sin(a) cos(b) = \frac{1}{2}[sin(a + b) + sin(a - b)]$:

$sin(x) cos(x) = \frac{1}{2}[sin(2x) + sin(0)] = \frac{1}{2} sin(2x)$, and now, to differentiate:

$f'(x) = \left(\frac{1}{2} sin(2x)\right)' = \frac{1}{2} \times 2 \times cos(2x) = cos(2x)$

Quotient Rule

- The Quotient Rule simplifies differentiation when dealing with fractions.
- The Quotient Rule is a formula used to find the derivative of a function that represents the quotient of two functions. Given two functions, $f(x)$ and $g(x)$, where $g(x)$ is not equal to *zero*, the Quotient Rule states that the derivative of their quotient, denoted as $\left(\frac{f(x)}{g(x)}\right)'$, is given by $\frac{(f'(x) \times g(x) - f(x) \times g'(x))}{(g(x))^2}$.
- In simple terms, to find the derivative of a quotient, we take the [numerator's derivative times the denominator], minus [the numerator times the denominator's derivative], all divided [by the square of the denominator].
- You can find the derivative of $\frac{1}{x}$ using this formula: $-\frac{1}{x^2}$.

This formula is a simple result of power rule of derivative. It can be found using quotient rule as well.

Examples:

Example 1. What's the derivative of $f(x) = \frac{3}{7x}$.
Solution: Using $\left(\frac{1}{x}\right)' = -\frac{1}{x^2}$, we have: $f'(x) = \left(\frac{3}{7x}\right)' = \frac{3}{7} \times \left(\frac{1}{x}\right)' = \frac{3}{7}\left(-\frac{1}{x^2}\right) = \frac{-3}{7x^2}$.

Example 2. Find the derivative of the function $f(x) = \frac{3x^2+1}{2x+1}$.
Solution: Using the Quotient Rule, we have:
$f'(x) = \frac{(3x^2+1)'(2x+1)-(2x+1)'(3x^2+1)}{(2x+1)^2} = \frac{6x(2x+1)-2(3x^2+1)}{(2x+1)^2} = \frac{12x^2+6x-6x^2-2}{4x^2+4x+1} = \frac{6x^2+6x-2}{4x^2+4x+1}$.

Example 3. Consider the function $g(x) = \frac{sin(x)+1}{cos(x)}$.
Solution: Applying the Quotient Rule, we obtain:
$g'(x) = \frac{(sin(x)+1)' cos(x)-(cos(x))'(sin(x)+1)}{(cos(x))^2} = \frac{cos(x)cos(x)-(-sin(x))(sin(x)+1)}{cos^2(x)}$
$= \frac{cos^2(x)-(sin^2(x)+sin(x))}{cos^2(x)} = \frac{cos^2(x)+sin^2(x)+sin(x)}{cos^2(x)} = \frac{(cos^2(x)+sin^2(x))+sin(x)}{cos^2(x)} = \frac{1+sin(x)}{cos^2(x)}$

Example 4. Suppose we have the function $h(x) = \frac{5x^3-2}{3x^2+1}$.
Solution: Using the Quotient Rule, we get:
$h'(x) = \frac{(5x^3-2)'(3x^2+1)-(3x^2+1)'(5x^3-2)}{(3x^2+1)^2} = \frac{15x^2(3x^2+1)-6x(5x^3-2)}{(3x^2+1)^2} = \frac{45x^4+15x^2-30x^4+12x}{9x^4+6x^2+1} = \frac{15x^4+15x^2+12x}{9x^4+6x^2+1}$

bit.ly/3GAm6Ux
Find more at

Chain Rule

- Chain rule has been explained earlier briefly. It can be shown using another format than: $f(x) = g(h(x)) \Rightarrow f'(x) = g'(h(x)) \times h'(x)$.

- If a function f is composed of two differentiable functions $y(x)$ and $u(x)$, so that $f(x) = y(u(x))$, then $f(x)$ is differentiable and: $\frac{df}{dx} = \frac{dy}{du} \times \frac{du}{dx}$.

- The method is called the "chain rule" because it can be applied sequentially to as many functions as are nested inside one another. For example, if f is a function of g which is in turn a function of h, which is in turn a function of x, that is $f(g(h(x)))$, the derivative of f with respect to x is given by: $\frac{df}{dx} = \frac{df}{dg} \times \frac{dg}{dh} \times \frac{dh}{dx}$.

- The Chain Rule's notation may suggest du "cancels out" in $\frac{dy}{du} \times \frac{du}{dx}$. That's not the case, because this is a mnemonic device, not a true cancellation. Differentials aren't numbers, but this perspective aids in recalling the rule's form.

Examples:

Example 1. If $u(x) = 3x^2 + 4x - 7$ and $y(u) = sin(u)$, find the derivative of f with respect to x: $f(x) = y(u(x))$.

Solution: By following the formula for chain rule, so:
$\frac{df}{dx} = \frac{dy}{du} \times \frac{du}{dx} = \frac{d}{du}(sin(u)) \times \frac{d}{dx}(3x^2 + 4x - 7) = cos(u)(6x + 4)$.
Since $u(x) = 3x^2 + 4x - 7$, in terms of x: $f'(x) = cos(3x^2 + 4x - 7)(6x + 4)$.

Example 2. Given the following functions, find the derivative of this composite function: $g(t) = f(y(u(x(t))))$, where $f(y) = 4y + 7$, $y(u) = u^3 + 2u^2 - 5$, $u(x) = 2x - 3$, and $x(t) = t^2 + 1$.

Solution: $g'(t) = \frac{df}{dy} \times \frac{dy}{du} \times \frac{du}{dx} \times \frac{dx}{dt}$, so: $g'(t) = 4 \times (3u^2 + 4u) \times 2 \times 2t$. Simplify:

$g'(t) = 16t(3u^2 + 4u)$. Now, substitute the function $u(x) = 2x - 3$ in the obtained equation. Then: $g'(t) = 16t(3(2x - 3)^2 + 4(2x - 3))$, and simplify:

$g'(t) = 16t(12x^2 - 28x + 15)$

Similarly, by substituting equation x in equation g' and simplifying, we have:

$g'(t) = 16t(12(t^2 + 1)^2 - 28(t^2 + 1) + 15) = 192t^5 - 64t^3 - 16t$.

So, the derivative of $g(t)$ with respect to t, is: $192t^5 - 64t^3 - 16t$.

Power Rule Combined with Other Derivative Rules

- The power rule can be combined with other derivative rules to find the derivative of more complex functions. For example, using the power rule in combination with the constant multiple rules, the derivative of a function $g(x) = cx^n$, where c is a constant, is given by $g'(x) = cnx^{n-1}$.

- Furthermore, the power rule can be extended to handle cases where the variable is not x but a function itself. This is accomplished using the chain rule. For instance, if we have a function $h(x) = (x^2 + 1)^3$, we can apply the power rule and chain rule together to find its derivative, by dealing with the exponent first $3(x^2 + 1)^2$, then multiplying this to $2x$, which is the derivative of the inner function.

Examples:

Example 1. Find the derivative of $f(x) = (2x^3 + 5x^2)^4$.

Solution: Using the power rule, we can differentiate the function as follows:
$$f'(x) = 4(2x^3 + 5x^2)^3 \times (6x^2 + 10x)$$

Here, we applied the power rule to the outermost function $(2x^3 + 5x^2)^4$, multiplying the exponent (4) with the expression $(2x^3 + 5x^2)^3$ like we normally do. Then differentiated the inner function using the power rule again which gives us $(6x^2 + 10x)$. So, when combined, we get: $f'(x) = 4(6x^2 + 10x)(2x^3 + 5x^2)^3$.

Example 2. Find the derivative of $f(x) = \frac{3x^4 - 2x^3}{x^2}$.

Solution: Using the power rule and the quotient rule, we differentiate the function as follows:
$$f'(x) = \frac{(12x^3 - 6x^2)(x^2) - (2x)(3x^4 - 2x^3)}{(x^2)^2}$$

Simplifying the expression, we get:
$$f'(x) = \frac{12x^5 - 6x^4 - (6x^4 - 4x^3)}{x^4} = \frac{6x^5 - 2x^4}{x^4} = 6x - 2$$

Therefore, the derivative of $f(x)$ is:
$$f'(x) = 6x - 2$$

Derivative of Radicals

- Derivative of $\sqrt{f(x)}$ can be calculated using $\frac{f'(x)}{2\sqrt{f(x)}}$ (so the derivative of $\sqrt{x}$ with index of 2 can be calculated from: $\frac{1}{2\sqrt{x}}$).

- For radicals with indexes other than 2, we can use: $\left(\sqrt[n]{x^m}\right)' = \frac{m}{n \times \sqrt[n]{x^{n-m}}}$

- Another way you can solve the radical derivatives, you can turn the $\sqrt{x^m}$ to $x^{\frac{m}{2}}$ or $\sqrt[n]{x^m}$ to $x^{\frac{m}{n}}$ and solve the derivative using power rule.

Examples:

Example 1. Find the derivative of $f(x) = \sqrt{x+2}$.

Solution: We can use $\frac{f'(x)}{2\sqrt{f(x)}}$ which gives: $f'(x) = \frac{1}{2\sqrt{x+2}}$. Also, we can see the function as $(x+2)^{\frac{1}{2}}$, and use power rule. So: $f'(x) = \frac{1}{2}(x+2)^{-\frac{1}{2}} \times 1 = \frac{1}{2\sqrt{x+2}}$.

Example 2. Find the derivative of $g(x) = \sqrt{3x^3}$.

Solution: We can change this to $(3x^3)^{\frac{1}{2}}$, so we have: $g'(x) = \frac{1}{2}(3x^3)^{-\frac{1}{2}} \times 9x^2$. Which is equal to $\frac{9x^2}{2\sqrt{3x^3}}$. We could reach the same result using: $\frac{f'(x)}{2\sqrt{f(x)}} = \frac{9x^2}{2\sqrt{3x^3}}$.

Example 3. Find the derivative of $h(x) = \sqrt{sin(5x)} + 4\tan^3(x)$.

Solution: We approach this problem step by step:

Step 1: $\left(\sqrt{sin(5x)}\right)' = \left((sin(5x))^{\frac{1}{2}}\right)' = \frac{1}{2}(sin(5x))^{-\frac{1}{2}} \times cos(5x) \times 5 = \frac{5\cos(5x)}{2\sqrt{sin(5x)}}$.

Step 2: $(4\tan^3(x))' = 4 \times 3 \times \tan^2(x) \times (1 + \tan^2(x)) = 12\tan^2(x) \times (1 + \tan^2(x))$.

So overall, we have: $\frac{5\cos(5x)}{2\sqrt{sin(5x)}} + \left(12\tan^2(x) \times (1 + \tan^2(x))\right)$.

Example 4. Solve $\left(\frac{\sqrt{3-x}}{(3x)^2}\right)'$.

Solution: We use quotient rule to find the derivative of $\frac{\sqrt{3-x}}{9x^2}$. So:

$$\left(\frac{\sqrt{3-x}}{9x^2}\right)' = \frac{(\sqrt{3-x})'9x^2 - (9x^2)'\sqrt{3-x}}{(9x^2)^2} = \frac{-\frac{1}{2\sqrt{3-x}}9x^2 - 18x\sqrt{3-x}}{81x^4}$$

$$= \frac{\frac{-9x^2 - 36x(3-x)}{2\sqrt{3-x}}}{81x^4} = \frac{27x^2 - 108x}{81x^4(2\sqrt{3-x})} = \frac{x-4}{6x^3\sqrt{3-x}}$$

Derivative of Logarithms and Exponential Functions

- For the derivative of logarithms, we have:
    1. **Derivative of $\log_a f(x)$** can be calculated using: $\frac{1}{f(x)\ln a} f'(x)$.
    2. **Derivative of $\log_a x$** can be calculated using: $\frac{1}{x \ln a}$, which is a special form of $(\log_a f(x))'$.
- For natural logarithms or ln, we have:
    1. **Derivative of $\ln f(x)$** can be calculated from: $\frac{1}{f(x)} f'(x)$.
    2. **Derivative of $\ln x$** is equal to $\frac{1}{x}$, which is a special form of $(\ln f(x))'$.
- For Exponential Functions, the derivatives can be calculated using:
    1. $(a^x)' = a^x \times \ln(a)$
    2. $\left(a^{f(x)}\right)' = a^{f(x)} \times \ln(a) \times f'(x)$
    3. $(e^x)' = e^x$
    4. $\left(e^{f(x)}\right)' = e^{f(x)} \times f'(x)$

Examples:

Example 1. Find the derivative of $\log_3 3x^2$.
Solution: Using $\log_a f(x) = \frac{1}{f(x)\ln a} f'(x)$: $\frac{1}{3x^2 \ln 3} 6x = \frac{2}{x \ln 3}$.

Example 2. Find the derivative of $\ln(x^2 + 1)$.
Solution: Using $\ln f(x) = \frac{1}{f(x)} f'(x)$, the function in parenthesis has a derivative of $2x$, which is multiplied to $\frac{1}{x^2+1}$, so the answer is: $\frac{2x}{x^2+1}$.

Example 3. Consider the function $g(x) = x \times \ln(x)$.
Solution: Using the Product Rule, we have:
$g'(x) = (1 \times \ln(x)) + \left(x \times \frac{1}{x}\right) = \ln(x) + 1$.

Example 4. $(2^x)'$.
Solution: $(a^x)' = a^x \times \ln(a) \Rightarrow (2^x)' = 2^x \times \ln(2)$.

Example 5. $(3^{4x})'$.
Solution: $\left(a^{f(x)}\right)' = a^{f(x)} \times \ln(a) \times f'(x) \Rightarrow (3^{4x})' = 3^{4x} \times \ln(3) \times 4$.

Example 6. $(e^2)'$.
Solution: We have a constant for exponent, and $e = 2.71 \cdots$ is a constant itself. So, the derivative of a number is zero.

L'Hôpital

- L'Hôpital's Rule is a technique used to evaluate limits of functions that are in an indeterminate form, such as $\frac{0}{0}$ or $\frac{\infty}{\infty}$.

- The rule states that if the limit of the quotient of two functions is indeterminate, taking the derivatives of the numerator and denominator separately and evaluating the limit of the resulting expression, can provide the correct result.

- This rule relies on the idea that <u>the behavior of a function near a point can be described by its derivative</u>. However, Hopital's Rule should be used with caution, ensuring that the functions are differentiable, and the limit is genuinely indeterminate, to obtain reliable results.

Examples:

Example 1. Let's find the limit of the function $f(x) = \frac{sin(x)-x^2}{x^2-7x}$ as x approaches $x = 0$.

Solution: We have an indeterminate form of $\frac{0}{0}$.

Applying Hopital's Rule, we take the derivatives of the numerator and denominator separately:

$y = sin(x) - x^2 \Rightarrow y' = cos(x) - 2x$

$y = x^2 - 7x \Rightarrow y' = 2x - 7$

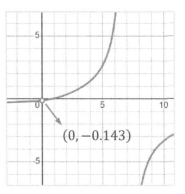

Now, we can evaluate the limit by plugging in $x = 0$ to the derived function:

$\lim\limits_{x \to 0} \frac{sin(x)-x^2}{x^2-7x} = \lim\limits_{x \to 0} \frac{cos(x)-2x}{2x-7} = \frac{cos(0)-2(0)}{2(0)-7} = \frac{1-0}{0-7} = -\frac{1}{7} = -0.143$

Example 2. Consider the limit of $g(x) = \frac{ln(x)}{x^2}$ as x approaches infinity ∞.

Solution: Again, we have an indeterminate form of $\frac{\infty}{\infty}$. Using Hopital's Rule, we take the derivatives of the numerator and denominator. Evaluating the limit as x approaches ∞: $\lim\limits_{x \to \infty} \frac{ln(x)}{x^2} = \lim\limits_{x \to \infty} \frac{\frac{1}{x}}{2x} = \lim\limits_{x \to \infty} \frac{1}{2x^2} = \frac{1}{\infty} = 0$.

Differentiability

- Differentiability is a fundamental concept in calculus that characterizes the smoothness of a function. A function is said to be differentiable at a point if it has a derivative at that point. In other words, it means the function can be approximated by a linear function (tangent line) at that specific point without any abrupt changes or discontinuities.

- A function is differentiable on an interval if it is differentiable at every point within that interval. So, differentiability implies continuity, but the converse is not always true. Non-differentiable points occur at corners, cusps, and points with vertical tangents. For example, the function $f(x) = |x|$ at $x = 0$.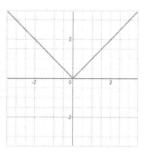

- Differentiability allows us to analyze how functions behave and make predictions about their behavior.

Examples:

Example 1. Consider the function $f(x) = x^2$, where $x \in \mathbb{R}$.

Solution: To determine if this function is differentiable, we need to examine its derivative. Taking the derivative of $f(x)$, we get $f'(x) = 2x$. Since this derivative exists for all real values of x, we can conclude that $f(x) = x^2$ is differentiable everywhere.

Example 2. Let's consider the function $g(x) = |x|$, where x represents any real number.

Solution: The absolute value function can be defined as: $g(x) = x$ for $x \geq 0$ and $g(x) = -x$ for $x < 0$. To determine the differentiability of $g(x)$, we need to examine the derivative. The derivative of $g(x)$ depends on the sign of x: $g'(x) = 1$ if $x > 0$, $g'(x) = -1$ if $x < 0$, and it is undefined at $x = 0$.

This function is continuous everywhere, including at $x = 0$. However, it is not differentiable at $x = 0$ because the slopes on either side of the point differ. The function has a sharp corner at $x = 0$, which prevents differentiability. So, the absolute value function, $f(x) = |x|$, does have a derivative, except at $x = 0$ where it is not differentiable.

Second Derivatives: Minimum vs. Maximum

- If we take the "derivative of the derivative" of a function, we took the second derivative of that function, denoted as $f''(x)$, so: $\left((f(x))'\right)' = f''(x)$.

- Any derivative beyond the first one is called a **higher order** derivative.

- The Second Derivative plays a crucial role in determining whether a function has a minimum or maximum at a given point.

- The second derivative essentially measures the rate of change of the first derivative. If the second derivative is positive, it indicates that the slope of the graph is increasing and thus the function is concave up. This suggests that there is a minimum at that point. Conversely, if the second derivative is negative, it implies that the slope of the graph is decreasing, and the function is concave down. This suggests that there is a maximum at that point.

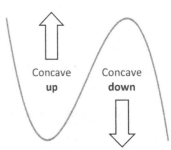
Concave up · Concave down

(Although sometimes points of inflection can occur where the second derivative is zero, also, the concavity may change without necessarily indicating a maximum or minimum.)

- To determine if a critical point corresponds to a minimum or maximum, we examine the sign of the second derivative at that point. Finding the critical points by setting the first derivative equal to *zero*.

Example: Let's consider the function $f(x) = x^3 - 3x^2 + 2x$. To find the second derivative, first we calculate the first derivative: $f'(x) = 3x^2 - 6x + 2 \Rightarrow$ second derivative: $f''(x) = 6x - 6$

$3x^2 - 6x + 2 = 0$, after solving for x, $x_1 = 1 - \frac{\sqrt{3}}{3} = 0.422$ and $x_2 = 1 + \frac{\sqrt{3}}{3} = 1.577$, Now let's evaluate the second derivative at these critical points: $f''\left(1 - \frac{\sqrt{3}}{3}\right) = -2\sqrt{3}$, which is less than zero. This indicates a local maximum at $1 - \frac{\sqrt{3}}{3}$, and $f''\left(1 + \frac{\sqrt{3}}{3}\right) = 2\sqrt{3}$, which is greater than zero. This indicates a local minimum at $1 + \frac{\sqrt{3}}{3}$. We can find the corresponding y values as:

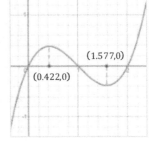

$f\left(1 - \frac{\sqrt{3}}{3}\right) = 0.3849$ and $f\left(1 + \frac{\sqrt{3}}{3}\right) = -0.3849$

Curve Sketching Using Derivatives

1. **Find the domain**: Determine the values of x for which the function is defined.

2. **Identify critical points**: These points occur when the derivative equals zero or is undefined. Solve for x to find these points.

3. **Determine the intervals of increase and decrease of $f(x)$**: Take derivative of the function (first derivative test) and evaluate the sign of the derivative in each interval. Positive values indicate increasing, while negative values indicate decreasing.

4. **Locate the local extrema**: Examine the critical points found in step 2 to determine if they correspond to local maxima or minima. You can do this by analyzing the behavior of the derivative around each critical point.

5. **Identify Points of Inflection**: Use the second derivative to identify intervals of positive and negative values. Positive values indicate concave up, while negative values indicate concave down.

6. **Sketch the curve**: Plot the critical points, local extrema, and inflection points on a graph and connect them smoothly, considering the behavior identified for each interval.

Example:

Given the function $f(x)$, find critical points, draw the graph and determine the intervals of increase and decrease $f(x) = x^3 - 3x^2 - 9x + 5$.

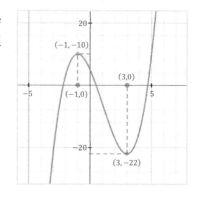

Solution: $f'(x) = 3x^2 - 6x - 9 \Rightarrow 3x^2 - 6x - 9 = 0$

Then: $(x - 3)(3x + 3) = 0 \Rightarrow x = 3$ and $x = -1$.

$(-\infty, -1)$: Increasing on this interval.

$(-1, 3)$: Decreasing on the interval.

$(3, +\infty)$: Increasing on the interval.

$f''(x) = 6x - 6 \Rightarrow 6x - 6 = 0 \Rightarrow x = 1$

The function is concave down on $(-\infty, 1)$, concave up on $(1, +\infty)$.

Differentiating Inverse Functions

- Differentiating inverse functions refers to the process of finding the derivative of a function that is the inverse of another function. In other words, if you have a function $f(x)$ and its inverse function $f^{-1}(x)$, differentiating the inverse function involves finding the derivative of $f^{-1}(x)$ with respect to x, or $(f^{-1})'(x)$.

- To find this, we first find the inverse of the function, then take its derivative.

- Inverse trigonometric functions like $\sin^{-1} x$ or $\arcsin x$, are functions that give the angle whose $\sin$ (or $\cos$, $\tan$ and $\cot$) is equal to a specified value. For example, if $\sin(90°) = 1$, then $\arcsin(1) = 90°$.

- To find their derivatives, we employ the following rules:

  - $(\arcsin(x))' = \frac{1}{\sqrt{1-x^2}}$
  - $(\arctan(x))' = \frac{1}{1+x^2}$
  - $(\arccos(x))' = \frac{-1}{\sqrt{1-x^2}}$
  - $(\text{arc}\cot(x))' = \frac{-1}{1+x^2}$

- Similarly, if we had $f(x)$ instead of x in the formulas above, we have:

  - $(\arcsin(f(x)))' = \frac{f'(x)}{\sqrt{1-(f(x))^2}}$
  - $(\arctan(f(x)))' = \frac{f'(x)}{1+(f(x))^2}$
  - $(\arccos(f(x)))' = \frac{-f'(x)}{\sqrt{1-(f(x))^2}}$
  - $(\text{arc}\cot(f(x)))' = \frac{-f'(x)}{1+(f(x))^2}$

Examples:

Example 1. Find the derivative of inverse of $y = 3x$.

Solution: To find the inverse, we switch x and y: $x = 3y$, so: $y = \frac{x}{3}$.

Please note that this new y is different from the original y. Now to take the derivative: $y' = \left(\frac{x}{3}\right)' = \frac{1}{3}$, and this is the derivative of inverse of $y = 3x$.

Example 2. Find the derivative of $\arcsin(3x)$.

Solution: We use the formula: $(\arcsin(f(x)))' = \frac{f'(x)}{\sqrt{1-(f(x))^2}}$.

Substituting the values, we have: $(\arcsin(3x))' = \frac{3}{\sqrt{1-(3x)^2}} = \frac{3}{\sqrt{1-9x^2}}$.

Please note that this is different from the inverse of $\sin(3x)$, which is $\frac{1}{3}\arcsin(x)$.

$\arcsin(3x)$ is an angle whose sine is three times x.

Optimization Problems

- Optimization problems in derivatives involve finding the maximum or minimum value of a function. These problems arise in various fields such as economics, engineering, and physics, where one seeks to optimize a certain quantity.

- The key strategy is to use derivatives to analyze the behavior of the function and identify critical points.

- To solve an optimization problem, we first define the objective function and identify the constraints. The objective function represents the quantity to be optimized, while the constraints represent the limitations or conditions that must be satisfied. We then take the derivative of the objective function and set it equal to zero to find critical points. These critical points may correspond to a maximum or minimum value of the function.

- Next, we evaluate the function at the critical points along with the endpoints of the domain to find the absolute maximum and minimum values. Additionally, we use the second derivative test to determine if these critical points are local maxima or minima.

Example:

Example 1. Suppose we have a function $f(x) = x^2 - 4x + 5$.

Solution: To find the minimum or maximum value of this function, we can take its derivative and set it equal to *zero*.

The derivative of $f(x)$ with respect to x is $f'(x) = 2x - 4$.

Setting $f'(x)$ equal to *zero*, we have $f'(x) = 0 \Rightarrow 2x - 4 = 0 \Rightarrow x = 2$.

To determine if this value of x corresponds to a minimum or maximum value, we can take the second derivative of $f(x)$. So, $f''(x) = 2$.

Since the second derivative is positive, we know that $x = 2$ corresponds to a minimum value of the function.

Thus, the minimum value of $f(x)$ is $f(2) = (2)^2 - 4(2) + 5 = 1$.

Implicit Differentiation

- Implicit differentiation is a method used to differentiate equations that involve both y and x without explicitly solving for y in terms of x. It's especially useful when trying to find the derivative of an equation where it's difficult or impossible to solve for y in terms of x. The method is based on chain rule.
- Here's how implicit differentiation works:
  1. **Differentiate Both Sides with Respect to x**: If you have an equation in the form $F(x,y) = 0$, then differentiating both sides with respect to x will give: $\frac{d}{dx}F(x,y) = \frac{d}{dx}0$.
  2. **Use the Chain Rule for y Terms**: When you encounter a y term during differentiation, use the chain rule. For example, if you differentiate a term like y^2, you treat y as an implicit function of x:
  $$\frac{d}{dx}y^2 = 2y\frac{dy}{dx}$$
  Notice how the term $\frac{dy}{dx}$ appears. This represents the derivative of y with respect to x, which is what we're trying to find:
  $\frac{d(y^2)}{dy} = 2y$, so $d(y)^2 = 2y\,dy$, now if we divide both by dx, we have:
  $$\frac{d(y^2)}{dx} = 2y\frac{dy}{dx}$$
  3. **Solve for $\frac{dy}{dx}$**: After differentiating both sides, you'll likely have an equation with $\frac{dy}{dx}$ terms. Now, you can solve for $\frac{dy}{dx}$ to find the derivative.

Examples:

Example 1. The equation $x^2 + y^2 = 1$ is given, (the unit circle equation), find $\frac{dy}{dx}$.
Solution: $\frac{d}{dx}(x^2 + y^2) = \frac{d(1)}{dx}$, from chain rule, we have: $2x + 2y\frac{dy}{dx} = 0$.
Now, solving for $\frac{dy}{dx}$, we have: $2y\frac{dy}{dx} = -2x$. After dividing both sides by $2y$: $\frac{dy}{dx}$, for the $x^2 + y^2 = 1$ circle is equal to: $\frac{dy}{dx} = -\frac{x}{y}$.

Example 2. Find $\frac{dy}{dx}$ for: $x^2y + xy^2 = sin(xy) + 3$.
Solution: We need to use the product rule because y is a function too, and after taking the derivative of both sides:
$\left(2xy + x^2\frac{dy}{dx}\right) + \left(y^2 + 2xy\frac{dy}{dx}\right) = cos(xy)\left(y + x\frac{dy}{dx}\right) + 0$. So:
$x^2\frac{dy}{dx} + 2xy\frac{dy}{dx} - x\,cos(xy)\frac{dy}{dx} = cos(xy)\,y - 2xy - y^2$. If we factor out the $\frac{dy}{dx}$:
$\frac{dy}{dx}(x^2 + 2xy - x\,cos(xy)) = cos(xy)\,y - 2xy - y^2$.
Finally, $\frac{dy}{dx} = \frac{cos(xy)y - 2xy - y^2}{x^2 + 2xy - x\,cos(xy)}$.

Related Rates

- Related rate is a concept in calculus that involves solving problems where multiple variables are changing with respect to time.

- These problems typically involve finding the rate at which one variable is changing when the rates of change of other related variables are known.

- In related rates problems, the chain rule plays a pivotal role in connecting the rates of change of interdependent quantities.

- To solve these problems, one typically uses calculus techniques such as differentiation and implicit differentiation to find the appropriate equations that relate the variables and their rates of change. Then, by taking the derivative with respect to time, the desired rate of change can be found.

- Geometric shapes (circles, rectangles, triangles) frequently appear in related rates problems, linking spatial changes to temporal ones.

- Real-world applications of related rates include physics scenarios like fluid flow, object movement, and shadow lengthening.

Example:

Air is pumped into a spherical balloon. As the radius increases at $0.1 \frac{ft}{s}$, how fast is the volume increasing when the radius is $2\,ft$?

Solution: Use the volume formula for spheres and differentiate with respect to time. We show volume as V, which is calculated using $V = \frac{4}{3}\pi r^3$.

Differentiate both sides with respect to time, t, to find the rate of change of volume: $\frac{dV}{dt} = \frac{4}{3} \times \pi \times 3 \times r^2 \frac{dr}{dt} = 4\pi r^2 \frac{dr}{dt}$.

And we know from that $\frac{dr}{dt}$ is $0.1 \frac{ft}{s}$, which is the rate at which the radius is increasing. Plugging in the given values: $\frac{dV}{dt} = 4\pi(2)^2(0.1) = 1.6\pi$.

So, when the radius is $2\,ft$, the volume is increasing at $1.6\pi \frac{ft^2}{s}$, (feet per second) or approximately $1.6\pi \approx 5.02$.

Chapter 5: Practices

✎ **Sketch the curve for the following functions using Derivatives.**

1) $f(x) = x^3$

2) $g(x) = 2x^2 - 8x + 10$

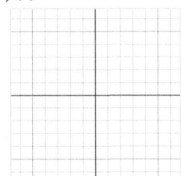
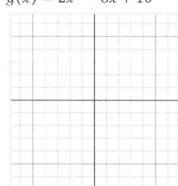

✎ **Determine differentiability.**

3) $f(x) = x^2 \Rightarrow$ ____

4) $g(x) = sin(x) \Rightarrow$ ____

5) $h(x) = |x| \Rightarrow$ ____

6) $I(x) = \frac{x^3+8}{4x} \Rightarrow$ ____

7) $j(x) = \sqrt{9x} \Rightarrow$ ____

8) $k(x) = \lfloor x^2 \rfloor \Rightarrow$ ____

9) $l(x) = \left(\frac{x}{2}\right)^3 - \left(\frac{x}{3}\right)^2 \Rightarrow$ ____

10) $m(x) = \frac{4x^2-12x+9}{(x-3)^3} \Rightarrow$ ____

✎ **Compute the derivative of the following functions by using derivative rules.**

11) $f(x) = x^{-2} \Rightarrow f'(x)$

12) $f(x) = 24 \Rightarrow f'(x)$

13) $f(x) = sin(3x^2) \Rightarrow f'(x)$

14) $f(x) = \frac{x+1}{x-1} \Rightarrow f'(x)$

15) $f(x) = x \cdot sin(x) \Rightarrow f'(x)$

16) $f(x) = \sqrt{x} - x \Rightarrow f'(x)$

17) $f(x) = x^{\frac{1}{2}} \Rightarrow f'(x)$

18) $f(x) = (3x + 1) \cdot x^3 \Rightarrow f'(x)$

Effortless Math Education

19) $f(x) = \frac{x^2}{sin(x)} \Rightarrow f'(x)$

20) $f(x) = (2x^2 + 4x + 1)^5 \Rightarrow f'(x)$

21) $f(x) = 2x^4 + 5x \Rightarrow f'(x)$

22) $f(x) = \sqrt{x} \cdot x^3 \Rightarrow f'(x)$

🖎 Solve.

23) $f(x) = ln(3x^2 + 4) \Rightarrow f'(x)$

24) $f(x) = log_3(x^2 + 1) \Rightarrow f'(x)$

25) $f(x) = x\, ln(x) \Rightarrow f'(x)$

26) $f(x) = \sqrt[3]{x^2 + 1} \Rightarrow f'(x)$

27) $f(x) = x\sqrt{x + 2} \Rightarrow f'(x)$

28) $f(x) = \frac{1}{\sqrt[4]{(x-3)}} \Rightarrow f'(x)$

29) $f(x) = (3x + 4) \times e^x \Rightarrow f'(x)$

30) $f(x) = sin^2(x) + cos(5x) \Rightarrow f'(x)$

🖎 Solve using L'Hôpital Rule.

31) $\lim\limits_{x \to 0} \frac{sin(x)}{x} = $ _____

32) $\lim\limits_{x \to 1} \frac{x^2 - 1}{x - 1} = $ _____

🖎 Find the derivative of inverse of these functions.

33) $f(x) = x^2 + 2$ for $x \geq 2$

34) $f(x) = 2^x$

35) $f(x) = sin(2x)$ for $\left[-\frac{\pi}{2}, \frac{\pi}{2}\right]$

36) $f(x) = 5x$

Chapter 5: Answers

1)
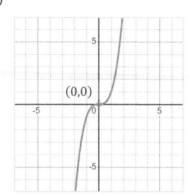

2)
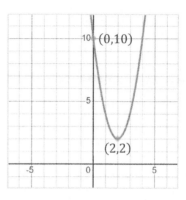

3) $f(x) = x^2$ is differentiable everywhere.

4) $g(x) = sin(x)$ is differentiable everywhere.

5) $h(x) = |x|$ is not differentiable at $x = 0$.

6) $I(x) = \frac{x^3+8}{4x}$ is differentiable everywhere except at $x = 0$.

7) $j(x) = \sqrt{9x}$ is differentiable for $x \geq 0$.

8) $k(x) = \lfloor x^2 \rfloor$ is not differentiable at integer values of x^2.

9) $l(x) = \left(\frac{x}{2}\right)^3 - \left(\frac{x}{3}\right)^2$ is differentiable for all real numbers.

10) $m(x) = \frac{4x^2-12x+9}{(x-3)^3}$ is differentiable for all real numbers except $x = 3$.

11) $-2x^{-3} = \frac{-2}{x^3}$

12) 0

13) $6x \cos(3x^2)$

14) $-\frac{2}{(x-1)^2}$

15) $x \cos(x) + \sin(x)$

16) $\frac{1}{2\sqrt{x}} - 1$

17) $\frac{1}{2} x^{-1/2} = \frac{1}{2\sqrt{x}}$

18) $3x^2(4x + 1)$

19) $\frac{2x \sin(x) - x^2 \cos(x)}{\sin^2(x)}$

20) $5(2x^2 + 4x + 1)^4(4x + 4)$

21) $8x^3 + 5$

22) $3.5\sqrt{x^5}$

Effortless Math Education

23) $\dfrac{6x}{3x^2+4}$

24) $\dfrac{2x}{(x^2+1)\ln(3)}$

25) $1 + \ln(x)$

26) $\dfrac{2x}{3(x^2+1)^{\frac{2}{3}}}$

27) $\sqrt{x+2} + \dfrac{x}{2\sqrt{x+2}}$

28) $\dfrac{-1}{4(x-3)^{\frac{5}{4}}}$

29) $(3e^x) + ((3x+4)e^x)$

30) $\sin(2x) - 5\sin(5x)$

31) 1

32) 2

33) $(f^{-1})'(x) = \dfrac{1}{2\sqrt{x-2}}$

34) $(f^{-1})'(x) = \dfrac{1}{x\ln 2}$

35) $(f^{-1})'(x) = \dfrac{1}{2\sqrt{1-x^2}}$

36) $(f^{-1})'(x) = \dfrac{1}{5}$

Chapter 6: Integrals

Math topics that you'll learn in this chapter:

- ☑ What is integral?
- ☑ Applications of integral
- ☑ Exponential growth and decay
- ☑ The anti-derivative
- ☑ Riemann sums
- ☑ Rules of Integration
- ☑ Power rule
- ☑ Fundamental theorem of calculus
- ☑ Trigonometric integrals
- ☑ Substitution rule
- ☑ Integration by parts
- ☑ Integral of radicals
- ☑ Exponential and logarithmic integrals
- ☑ Improper integrals

What is Integral?

- Integral is a fundamental concept in calculus that measures the accumulation of a quantity over an interval.

- It is denoted by the symbol $\int$ and computed through the integration process.

- Integrals have two main types: definite and indefinite integrals.

- Integral represents the area under a curve or the total of something. By finding the integral of a function, we can determine the exact value of the accumulated quantity, or the total change of a given variable.

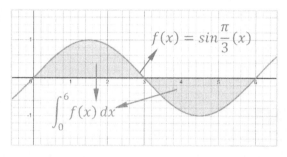

- The concept of numerical integration is the process of approximating the area under a curve by dividing the interval into smaller segments and summing up these approximations. As the number of subdivisions increases, the accuracy of the estimate improves, eventually yielding the exact area value.

- Numerical integration encompasses various techniques to approximate definite integrals. It includes methods like the trapezoidal rule, Simpson's rule, and Gaussian quadrature, among others. Numerical integration aims to provide accurate estimates of integrals when an analytical solution is difficult to obtain.

- Riemann sums is a basic method within numerical integration that breaks the curve into rectangles for approximation.

- Integrals have various applications in mathematics, such as determining distances traveled, finding areas of irregular shapes, and solving differential equations. They play a crucial role in understanding and solving real-world problems using mathematical tools.

- Integrals are the opposite of derivatives. Meaning that if we take the derivative of the answer to an integral, we get the expression inside the integral. For example: $\int x\, dx = \frac{x^2}{2}$ and $\left(\frac{x^2}{2}\right)' = \frac{1}{2} \cdot 2x = \frac{2x}{2} = x$.

Applications of Integrals

- **Calculating area**: Integrals can be used to calculate the area between a curve and the x-axis or between two curves.

- **Finding volumes**: Integrals can be used to calculate the volume of shapes that can be obtained by rotating a curve around an axis. For instance, finding the volume of a solid obtained by rotating a curve around the $x-axis$ can be done by using the method of disks or the method of cylindrical shells.

- **Calculating work**: Integrals are used to calculate work done in physics problems. For example, the work done by a varying force can be found by integrating the force over the displacement.

- **Computing probabilities**: Integrals are used to compute probabilities in statistics and probability theory. For example, finding the probability of an event occurring within a certain range can be done by integrating the probability density function over that range.

- **Calculating arc length**: Integrals are used to calculate the length of a curve. For instance, finding the length of a curve defined by a function can be done using: $L = \int_a^b \sqrt{1 + \left(\frac{dy}{dx}\right)^2}\, dx$.

- **Calculating centroids**: Integrals are used to calculate the coordinates of the center of mass or centroid of an object. For instance, finding the centroid of a region with variable density can be done by integrating the product of the density and the coordinates.

- **Finding remaining reactants**: In chemistry, integrals are applied to determine the total quantity of substances involved in chemical reactions over time, aiding in the analysis of reaction kinetics and stoichiometry.

- **Calculating Heat Transfer**: Integrals are used to calculate heat transfer in thermodynamics and engineering. For example, finding the amount of heat transferred in a process can be done by integrating temperature changes over time.

- **Analyzing Financial Risk**: Integrals can be applied to calculate the risk associated with financial investments by modeling the probability distribution of returns and finding expected values through integration techniques.

Exponential Growth and Decay

- Exponential growth refers to a situation where the integral of an exponentially increasing function grows rapidly over time.

- It occurs when the rate of change of a quantity is proportional to the current value of that quantity. As the function continues to grow exponentially, so does its integral.

- This type of growth is often observed in various fields, such as population growth, finance, and physics.

- Exponential growth in integrals can lead to significant accumulations or expansions over time and is mathematically represented by functions like the exponential function or its variants.

- Exponential decay refers to the mathematical concept where a function decreases exponentially over time or distance, and the integral of that function measures the accumulated decay.

- The integral of the exponential decay function yields the total amount of decay or the accumulated loss of a quantity.

- Exponential decay often occurs in natural phenomena like radioactive decay, population growth, or electrical circuits.

- By evaluating the integral of the exponential decay function, one can understand the extent of the decay or the quantity that remains after a certain period or distance.

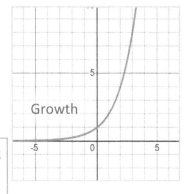

Growth

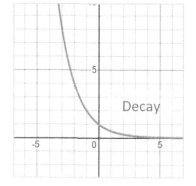

Decay

bit.ly/41gEviz

The Anti-Derivative

- The integral can be thought of as the inverse operation of differentiation or anti-derivative. When we differentiate a function, we're looking at how it changes at any given point, essentially obtaining its rate. Conversely, when we integrate, we're trying to determine the total accumulation based on a given rate of change.

- If we don't bound the integral by a particular interval, we have indefinite integrals, denoted as: $\int f(x)\,dx = F(x) + C$.
  Where $\int$ denotes integration, $f(x)$ is the integrand (the function to be integrated), dx represents the differential element, $F(x)$ is the antiderivative of $f(x)$, and C is an arbitrary constant.
  So, if $\int f(x)\,dx = F(x)$, then $F'(x) = f(x)$.

- The result of an indefinite integral always includes an arbitrary constant C, since the process of differentiation eliminates constants. Thus, when integrating $f(x)$, you get an anti-derivative plus C. This is because:
  Suppose $f(x) = 2x^3 + 4x^2 + 6x + 8$ and $g(x) = 2x^3 + 4x^2 + 6x + 1$. If we compare the derivatives of these functions, we see both functions have the derivative of $h(x) = 6x^2 + 8x + 6$, even though we clearly have two different functions. To find the integral of the function $h(x)$, we have to account for this, which we do by adding C:
  $\int (6x^2 + 8x + 6)\,dx = 2x^3 + 4x^2 + 6x + C$, where C can be 1, 8, 0 or any other number.

- When finding the area under a curve between the interval of two bounds $[a, b]$, we use: $\int_a^b f(x)\,dx$, the integral notation is modified by specifying the bounds of integration, and is called "**Definite integral**".

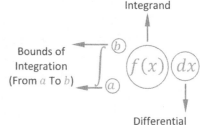

Example:

Example 1. Find the answer to this integral, based on the definition of anti-derivative: $\int (5x - 2)\,dx$.

Solution: To find the integral, we need to find an expression that when we take derivative of, results in $5x - 2$. From power rule, we know that x's exponent goes to be the new coefficient, and the exponent gets decreases by 1. So:
$\int 5x\,dx = \frac{5x^2}{2} + C = 2.5x^2 + C$. So, if applying power rule: $2 \times 2.5 \times x = 5x$, and $\int -2\,dx = -2x + C$. So, the original expression, meaning the answer to our integral, is: $\int (5x - 2)\,dx = 2.5x^2 - 2x + c$.

Riemann Sums

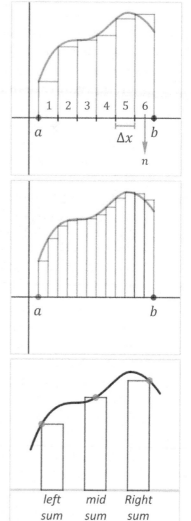

- Riemann Sums serve as a method to approximate the definite integral of a function over a specified interval.
- They work by breaking down the region under the curve of a function into simpler shapes, such as rectangles, that are easier to compute. The more rectangles you use (i.e., the narrower they are), the closer the approximation gets to the actual integral value.
- We begin by dividing the interval $[a,b]$ on which you want to approximate the integral into n subintervals of equal Δx width. Thus, $\Delta x = \frac{b-a}{n}$. The endpoints of these subintervals are $(x_1, x_2, \cdots, x_n)$, where $(x_1 = a)$ and $(x_n = b)$.
- For each subinterval, we choose a sample point. For **Left Riemann Sum**, the function value at the left endpoint will determine the height of the rectangle. Similarly, for **Right Riemann Sum**, we use the function value at the right endpoint. For **Midpoint Riemann Sum**, we use the function value at the midpoint of the subinterval.
- For each subinterval, form a rectangle with base Δx and height $f(x_i)$, the area of this rectangle is $f(x_i)\Delta x$. Sum the areas of all such rectangles to get the Riemann Sum: $S_n = \sum_{i=1}^{n} \Delta x f(x_i)$. As the number of subintervals (n) approaches infinity (and thus Δx approaches 0), the Riemann Sum approaches the definite integral of $f(x)$ over $[a,b]$: $\lim\limits_{n \to \infty} S_n = \int_a^b f(x)dx$.

Example:

Approximate the integral of the function $f(x) = x^2$ from $x = 0$ to $x = 2$ using the Left Riemann Sum using four subintervals.

Solution: Divide the [0,2] into 4 equal subintervals: [0,0.5], [0.5,1], [1,1.5], [1.5,2] Width of each subinterval is $\Delta x = \frac{2-0}{4} = 0.5$, now to calculate the left Riemann sum:

$$(f(0) \times 0.5) + (f(0.5) \times 0.5) + (f(1) \times 0.5) + (f(1.5) \times 0.5)$$
$$= (0^2 \times 0.5) + (0.5^2 \times 0.5) + (1^2 \times 0.5) + (1.5^2 \times 0.5)$$
$$= 0 + 0.125 + 0.5 + 1.125 = 1.75$$

Rules of Integration

- **Power Rule**: The integral of a power function with respect to the variable is given by: $\int x^n \, dx = \frac{x^{n+1}}{n+1} + C$, where n is not equal to -1.

- **Multiplication by constant Rule**: this rule states that when integrating a constant multiple of a function, the constant can be brought outside the integral sign. $\int c \cdot f(x) \, dx = c \int f(x) \, dx$, for example:

$$\int 5x \, dx = 5 \int x \, dx = 5 \frac{x^2}{2} + C$$

- **Linearity Rule**: The integral of a sum or difference of functions is the sum or difference of their integrals: $\int (f(x) \pm g(x)) \, dx = \int f(x) \, dx \pm \int g(x) \, dx$. For example: $\int (3x^2 + 4x) \, dx = \int 3x^2 \, dx + \int 4x \, dx$.

- **Substitution Rule**: This technique involves replacing a variable with a new variable or function to simplify the integral. Suppose we have an integral in the form $\int f(g(x)) \cdot g'(x) \, dx$, then we let $u = g(x)$ and $du = g'(x) \, dx$, leading to: $\int f(g(x)) \cdot g'(x) \, dx = \int f(u) \, du$.

- **Integration by Parts**: This rule is a counterpart to the product rule for differentiation. It states that if we have two functions, $u(x)$ and $v'(x)$, then the integral of their product is given by:

$$\int u(x) v'(x) \, dx = u(x) v(x) - \int v(x) u'(x) \, dx$$

- **Trigonometric Integrals**: Various trigonometric functions have specific integral formulas, such as:

  - $\int \sin(x) \, dx = -\cos(x) + C$
  - $\int \cos(x) \, dx = \sin(x) + C$

- **Exponential and Logarithmic Integrals**: Integrals involving exponential and logarithmic functions have specific rules and solutions based on their properties.

- **Radical Integrals**: solutions involving integrals can get quite complex, but for simple radical forms, there are formulas to easily find the integral:

  - $\int \sqrt{x} \, dx = \frac{2}{3} x^{\frac{3}{2}} + C$
  - $\int \sqrt{nx^m} \, dx = \frac{2\sqrt{n}\sqrt{x^{m+2}}}{m+2} + C$

bit.ly/41cGlf6

Power Rule

- The Power Rule of Integration states that for any function of the form $f(x) = x^n$, where n is a constant ($\in \mathbb{R}$), the integral of $f(x)$ with respect to x is given by: $\int (x^n)\, dx = \left(\frac{1}{n+1}\right) \cdot x^{n+1} + C$.

- In simple terms, if we have a function that is a power of x, such as x^2 or x^3, we can find its integral by adding 1 to the exponent, dividing the result by the new exponent. The resulting term is then added to the constant of integration.

- Needless to say, if x had a coefficient, multiply it by the result of integration, before adding C. (Multiplication by constant rule)

- To find the integral of $\frac{1}{x}$, we can't use this power rule, because we encounter $\frac{x^0}{0} + C$, instead we use this formula: $\int \frac{1}{x}\, dx = ln|x| + C$.

$\int \frac{1}{f(x)}\, dx$ doesn't have a general formula and depends on what function the $f(x)$ is.

Examples:

Example 1. Integrate the function $f(x) = x^3$.

Solution: $\int x^3\, dx = \frac{1}{3+1} \cdot x^{3+1} + C = \frac{x^4}{4} + C$.

Therefore, the indefinite integral of $f(x) = x^3$ is: $\frac{x^4}{4} + C$.

Example 2. Integrate the function $g(x) = 3x^2$.

Solution: $\int (3x^2)\, dx = 3 \times \left(\frac{1}{2+1}\right) \cdot x^{2+1} + C = 3 \times \left(\frac{1}{3}\right) \cdot x^3 + C = x^3 + C$.

So, the integral of $3x^2$ with respect to x is $x^3 + C$.

Example 3. Integrate the function $h(x) = x^{-3}$.

Solution: $\int x^{-3}\, dx = \frac{1}{-3+1} \cdot x^{-3+1} + C = \frac{x^{-2}}{-2} + C = \frac{-1}{2x^2} + C$.

Example 4. Integrate the function $I(x) = -4x^{-5}$.

Solution: $\int -4x^{-5}\, dx = \frac{1}{-5+1} \times x^{-5+1} \times (-4) + C = -4 \times \frac{-1}{4} x^{-4} + C = x^{-4} + C$.

We can always check the answer to an integral, by differentiating the result to get the original function: $(x^{-4})' = -4x^{-5}$.

Fundamental Theorem of Calculus

- The Fundamental Theorem of Calculus (FTC) is a foundational result in calculus that intimately connects the two main branches of the subject: differentiation and integration.
- The theorem consists of two parts, each of which ties the concept of derivatives to integrals.
    1. **First Part (FTC Part I):**
        - Given a continuous function $f(x)$ on an interval $[a, b]$, if F is an antiderivative of f on $[a, b]$, then: $\int_a^b f(x)\, dx = F(b) - F(a)$.
- This result provides a way to evaluate <u>definite integrals</u> without having to use Riemann sums or limit processes. Simply find antiderivative (F) of f, and compute the difference in F's values at the endpoints of the interval.
    2. **Second Part (FTC Part II):**
        - If f is a function that is continuous over an interval $[a, b]$ and F is defined on $[a, b]$ by: $F(x) = \int_a^x f(t)\, dt$, then F is differentiable on the open interval (a, b), and $F'(x) = f(x)$ for all x in (a, b).
- This result effectively says that the derivative of the integral of a function is the original function, which was mentioned earlier.
- The Fundamental Theorem of Calculus often uses the notation of "indefinite integral in brackets with lower and upper bounds as subscripts and superscripts": $\int_a^x f(t)\, dt = [F(t)]_a^x = F(x) - F(a)$, signifying the evaluation of the indefinite integral from a to x.

Example: Find the exact value $\int_1^3 f(x)\, dx = \int_1^3 x^2\, dx$, using Fundamental Theorem of Calculus.

Solution: First, we find the Antiderivative:
$F(x) = \int f(x)\, dx = \int x^2\, dx = \frac{1}{3}x^3 + C$.

The exact value of C is not relevant, as <u>it will cancel out</u> when we evaluate at our limits of integration.

Apply the FTC Part I: the definite integral from 1 to 3: $\int_1^3 x^2\, dx$

$[F(x)]_1^3 = [\frac{x^3}{3}]_1^3 = F(3) - F(1) = \left[F(3) = \frac{1}{3}(3)^3 = \frac{1}{3}(27) = 9\right]$ and $\left[F(1) = \frac{1}{3}(1)^3 = \frac{1}{3}(1) = \frac{1}{3}\right] \Rightarrow \int_1^3 x^2\, dx = 9 - \frac{1}{3} = \frac{26}{3}$.

Trigonometric Integrals

- Trigonometric Integrals refer to integrals that involve trigonometric functions.
- Solving these often requires recognizing certain patterns, identities, or combinations of sine, cosine, tangent, and other trigonometric functions.
- Familiarity with basic trigonometric identities, such as the Pythagorean identities or double-angle formulas, can be particularly helpful. For instance, converting powers of sine and cosine into a single trigonometric function or using the identity $sin^2(x) + cos^2(x) = 1$ can simplify many integrands, making the integration more tractable.
- Here are some of the basic formulas and techniques for trigonometric integrals:

$\int sin(x)\, dx = -cos(x) + C$ $\qquad$ $\int cos(x)\, dx = sin(x) + C$
$\int tan(x)\, dx = -ln|cos(x)| + C$ $\qquad$ $\int cot(x)\, dx = ln|sin(x)| + C$

- More advanced versions of these formulas, are:

$\int sin(nx)\, dx = -\frac{1}{n} cos(nx) + C$ $\qquad$ $\int cos(nx)\, dx = \frac{1}{n} sin(nx)$
$\int tan(nx)\, dx = -\frac{1}{n} ln|cos(nx)| + C$ $\qquad$ $\int cot(x)\, dx = \frac{1}{n} ln|sin(nx)| + C$

Examples:

Example 1. Find $\int sin(x) cos(x)\, dx$.

Solution: We know that $sin(2x) = 2 sin(x) cos(x)$, thus, our integral can be rewritten as $\frac{1}{2} \int sin(2x)\, dx$. Now, integrating with respect to x, we get:

$$\frac{1}{2}\left(-\frac{1}{2} cos(2x)\right) + C = -\frac{1}{4} cos(2x) + C$$

Example 2. $\int tan(x) cot(x)\, dx$.

Solution: Using the identity $tan(x) cot(x) = 1$, the integral becomes

$$\int 1\, dx = x + C$$

Example 3. $\int cos^2(x) tan(x)\, dx$.

Solution: We know that $tan(x) = \frac{sin(x)}{cos(x)}$, so simplifying the integral, we have:

$\int sin(x) cos(x)\, dx = -\frac{1}{4} cos(2x) + C.$

Substitution Rule

- The Substitution Rule for integration, often referred to as $u-$substitution, provides a method to transform a complex integral into a simpler one.

- Essentially, it involves introducing a new variable, often denoted as u, by relating it to a portion of the integrand or the inner function. Once this substitution is made, the differential du is also determined, allowing the original integral to be expressed entirely in terms of u. The integral is then solved in this new variable, and the result is reverted to the original variable (x) to provide the final solution.

- In practice, the method resembles the chain rule for differentiation but works in reverse.

- If one is faced with the integral of a composite function like $\int f(g(x)) \cdot g'(x)\, dx$, the substitution $u = g(x)$ simplifies this to $\int f(u)\, du$.

- This approach is particularly useful for integrands that have nested or composite structures.

Examples:

Example 1. Evaluate $\int 2x\, sin(x^2 + 1)\, dx$.

Solution: Looking at the integrand, we notice that the derivative of $x^2 + 1$ is $2x$, which is present as a factor. This observation leads us to try a substitution. Let: $u = x^2 + 1$. Then: $\frac{du}{dx} = 2x \Rightarrow du = 2x\, dx$.

With this substitution, our integral transforms to: $\int sin(u)\, du$.

Now, integrating with respect to u: $\int sin(u)\, du = -cos(u) + C$.

Re-substitute to express the result in terms of x: $-cos(x^2 + 1) + C$.

Therefore: $\int (2x \cdot sin(x^2 + 1))\, dx = -cos(x^2 + 1) + C$.

Example 2. Evaluate $\int (3x^3 + 2x)\, dx$ using substitution rule.

Solution: Rewrite the integrand as $\int x(3x^2 + 2)\, dx$. Let: $u = 3x^2 + 2$, then: $\frac{du}{dx} = 6x \Rightarrow \frac{du}{6} = x\, dx$. So, the new integral is: $\int \frac{u}{6}\, du = \frac{u^2}{12}$, and when replacing x:

$$\frac{u^2}{12} = \frac{(3x^2 + 2)^2}{12} = \frac{9x^4 + 12x^2 + 4}{12} = \frac{3}{4}x^4 + x^2 + \frac{1}{3} + C$$

Integration by Parts

- Integration by Parts is a fundamental technique in integral calculus, derived directly from the product rule for differentiation. The formula for integration by parts is given by: $\int u\,dv = uv - \int v\,du$.

  Where u and dv are chosen parts of the original integrand. The selection of these parts often relies on strategic choices to simplify the resulting integral.

- Once u and dv are identified, their corresponding derivatives, du and integrals, v are determined.

Examples:

Example 1. $\int x^2 \cdot x\,dx$.

Solution: $u = x^2 \Rightarrow \frac{du}{dx} = 2x \Rightarrow du = 2x\,dx$ and $dv = x\,dx$, integrating both sides: $v = \frac{1}{2}x^2$. Using the formula $\int u\,dv = uv - \int v\,du$:

$$\int x^2 \cdot x\,dx = x^2 \times \frac{1}{2}x^2 - \int \frac{1}{2}x^2 \cdot 2x\,dx = \frac{1}{2}x^4 - \int x^3\,dx = \frac{1}{2}x^4 - \frac{1}{4}x^4 = \frac{1}{4}x^4 + C$$

Note: In this specific case, it's more straightforward to just multiply $x^2 \cdot x = x^3$ and then integrate, but the process above aims to demonstrate the method of integration by parts.

Example 2. $\int x\sin(x)\,dx$.

Solution: If we choose $u = x$, which means $du = dx$, and $dv = \sin(x)\,dx$, which gives $v = -\cos(x)$. After applying the formula $\int u\,dv = uv - \int v\,du$, we have:

$$\int x\sin(x)\,dx = -x\cos(x) - \int(-\cos(x))\,dx = -x\cos(x) + \int \cos(x)\,dx$$
$$= -x\cos(x) + \sin(x) + C$$

Example 3. $\int x\cos(2x)\,dx$.

Solution: We choose $u = x$, so $du = dx$, and $dv = \cos(2x)\,dx$, which means: $v = \frac{1}{2}\sin(2x)$, and if we use the formula for integration by parts:

$$\int x\cos(2x)\,dx = \left(\frac{x}{2}\sin(2x) - \int \frac{1}{2}\sin(2x)\,dx\right) = \frac{x}{2}\sin(2x) + \frac{1}{4}\cos(2x) + C$$

Integral of Radicals

- The integral of $\sqrt{x}$ can be calculated using the following formula, which is a direct application of the power rule for integration:
$$\int \sqrt{x}\, dx = \frac{2}{3}\sqrt{x^3} + C$$

- If x had a coefficient, the formula changes to:
$$\int \sqrt{nx}\, dx = \frac{2\sqrt{n}}{3}\sqrt{x^3} + C$$

- In case of having both exponent and coefficient:
$$\int \sqrt{nx^m}\, dx = \frac{2\sqrt{n}\sqrt{x^{m+2}}}{m+2} + C$$

- If the radical had indexes other than 2, the formula is:
$$\int \sqrt[t]{nx^m}\, dx = \frac{n^{\frac{1}{t}}}{\left(\frac{m}{t}+1\right)} \times x^{\left(\frac{m}{t}+1\right)} + C$$

Examples:

Example 1. $\int \sqrt{5x}\, dx$.

Solution: You can solve this using power rule, or the formulas mentioned above. Using power rule. We can simplify $\sqrt{5x}$ to $\left(\sqrt{5} x^{\frac{1}{2}}\right)$, now to integrate this: $\int \sqrt{5} x^{\frac{1}{2}}\, dx = \sqrt{5} \times \frac{x^{\frac{3}{2}}}{\frac{3}{2}} = \frac{2\sqrt{5}}{3} x^{\frac{3}{2}} + C$. We obtain the same result using $\int \sqrt{nx}\, dx = \frac{2\sqrt{n}}{3}\sqrt{x^3} + C$: $\int \sqrt{5x}\, dx = \frac{2\sqrt{5}}{3}\sqrt{x^3} + C$.

Example 2. $\int x\sqrt{x}\, dx$.

Solution: $\int \sqrt{x^2}\sqrt{x}\, dx = \int \sqrt{x^3}\, dx = \int x^{\frac{3}{2}}\, dx = \frac{x^{\frac{3}{2}+1}}{\frac{3}{2}+1} + C = \frac{2}{5} x^{\frac{5}{2}} + C$.

Example 3. Find the integral $\int \frac{\sqrt{x}}{1+\sqrt{x}}\, dx$.

Solution: At first glance, the integrand looks quite complex. However, with a suitable substitution, it simplifies significantly. Let $u = 1 + \sqrt{x}$, so: $\sqrt{x} = u - 1$. If we differentiate both sides with respect to x to get du: $du = \frac{1}{2\sqrt{x}} dx$ or $dx = 2(u-1)\, du$. Substituting values: $\int \frac{\sqrt{x}}{1+\sqrt{x}}\, dx = \int \frac{u-1}{u} 2(u-1)\, du = 2\int \frac{u^2+1-2u}{u}\, du$
$= 2\int u + \frac{1}{u} - 2\, du = 2\left(\frac{u^2}{2} + \ln|u| - 2u + C\right) = u^2 + 2\ln|u| - 4u + C$.

Now to substitute $u = 1 + \sqrt{x}$ back into the expression:
$u^2 + 2\ln|u| - 4u + C = \left(1+\sqrt{x}\right)^2 + 2\ln\left(1+\sqrt{x}\right) - 4\left(1+\sqrt{x}\right) + C$.

Exponential and Logarithmic Integrals

- Exponential and Logarithmic Integrals are integrals involving exponential functions, such as e^x, or logarithmic functions, such as $ln(x)$.

- For exponential functions, the integral of e^x with respect to x is straightforward, as the antiderivative remains an exponential function of the same form. On the other hand, logarithmic functions present their own challenges. The integral of $ln(x)$ with respect to x, for example, requires integration by parts to evaluate.

- Exponential and logarithmic functions are closely related, especially considering the logarithm is the inverse of the exponential function. As such, certain integrals may require recognizing and converting between the two forms for simplification. Some basic formulas include:

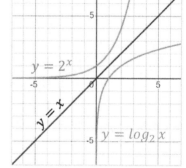

- $\int e^{ax} dx = \frac{1}{a} e^{ax} + C$
- $\int ln(x) dx = x(ln(x) - 1) + C$

Examples:

Example 1. Find $\int e^{3x} dx$.

Solution: Using $\int e^{ax} dx = \frac{1}{a} e^{ax} + C$, then: $\int e^{3x} dx = \frac{1}{3} e^{3x} + C$.

Example 2. Find the integral: $\int x e^{2x} dx$.

Solution: Using integration by parts, if $u = x$ and $dv = e^{2x} dx$, which means $v = \frac{1}{2} e^{2x}$, using $\int u \, dv = uv - \int v \, du$:

$\int x e^{2x} dx = x \left(\frac{1}{2} e^{2x}\right) - \int \frac{1}{2} e^{2x} dx = \frac{1}{2} x e^{2x} - \left(\frac{1}{2}\right)\left(\frac{1}{2}\right)(e^{2x}) + C = \frac{1}{2} x e^{2x} - \frac{1}{4} e^{2x} + C$.

Example 3. Calculate $\int ln(x) dx$.

Solution: Let $u = ln \, x$ and $dv = dx$, differentiating both sides: $\frac{du}{dx} = \frac{1}{x}$, meaning $du = \frac{dx}{x}$ and since $dv = dx$, then $v = x$, so we have:

Integration by parts: $x \, ln(x) - \int x \times \frac{1}{x} dx = x \, ln(x) - x + C$.

Improper Integrals

- Improper Integrals arise when one tries to compute the definite integral of a function over an interval where the function is unbounded (has infinite discontinuities) or the interval itself is unbounded.

- Essentially, they represent scenarios where the "area" being calculated might be infinite or where the region of integration extends indefinitely.

- There are two primary types of improper integrals:

    I. Infinite Intervals   II. Discontinuous Integrands

    - **Infinite Intervals** occur when the interval of integration is infinite. For example: $\int_1^\infty f(x)\,dx$, to handle such an integral, a limit is employed:
    $$\int_1^\infty f(x)\,dx = \lim_{b \to \infty} \int_1^b f(x)\,dx$$

    - **Discontinuous Integrands** happen when the function being integrated has a discontinuity on the interval of integration. For instance, consider: $\int_0^1 \frac{1}{\sqrt{x}}\,dx$, the function $\frac{1}{\sqrt{x}}$ is not defined at $x = 0$. To compute this integral, a limit is used: $\lim_{a \to 0^+} \int_a^1 \frac{1}{\sqrt{x}}\,dx$.

- For both types, the improper integral is considered convergent if the corresponding limit exists and is finite; Otherwise, it's divergent. So, the process of evaluating an improper integral doesn't guarantee a finite result. If the integral converges, then it has a finite value. If it diverges, the "area" being calculated is infinite.

Example:

Solve: $\int_1^\infty \frac{1}{x^2}\,dx$.

Solution: $\lim_{b \to \infty} \int_1^b \frac{1}{x^2}\,dx \Rightarrow$ Integrate $\frac{1}{x^2}\,dx$: $\int x^{-2}\,dx = -x^{-1} = -\frac{1}{x} + C$.

Now, evaluate this antiderivative from 1 to b: $F(b) - F(1) = -\frac{1}{b} - \left(-\frac{1}{1}\right) = 1 - \frac{1}{b}$.

$$\lim_{b \to \infty} 1 - \frac{1}{b} = 1 \Rightarrow \int_1^\infty \frac{1}{x^2}\,dx = 1$$

Chapter 6 Practices

✎ Solve using the Rules of Integration.

1) $\int x^{-3}\, dx = $ _____

2) $\int \frac{1}{5+3x}\, dx = $ _____

3) $\int x^2 e^{3x}\, dx = $ _____

4) $\int \sqrt{2x+1}\, dx = $ _____

5) $\int \sqrt{x} = $ _____

6) $\int \frac{dx}{\sqrt{x+4}} = $ _____

7) $\int x \cos(x)\, dx = $ _____

8) $\int \ln(x)\, dx = $ _____

✎ Solve.

9) $\int_0^2 x^2\, dx = $ _____

10) $\int_0^\pi \sin(x)\, dx = $ _____

11) $\int_1^e \ln(x)\, dx = $ _____

12) $\int_0^{\frac{\pi}{4}} \tan(x)\, dx = $ _____

✎ Approximate using left Riemann sums using given number of subintervals.

13) $\int_0^1 x^2\, dx \qquad n = 2$

14) $\int_1^3 x^3 - 2x^2 + 1\, dx \quad n = 4$

15) $\int_0^2 \sqrt{4-x^2}\, dx \qquad n = 6$

16) $\int_0^4 x - x^2\, dx \qquad n = 4$

✎ Solve the Improper Integrals.

17) $\int_0^1 \frac{1}{\sqrt{x}}\, dx = $ _____

18) $\int_0^\infty e^{-x}\, dx = $ _____

19) $\int_1^\infty \frac{1}{x}\, dx = $ _____

20) $\int_0^\infty 3x^3\, dx = $ _____

✎ Solve using the Fundamental Theorem of Calculus.

21) $\int_0^2 (2x^3 + 3x^2)\, dx = $ _____

22) $\int_0^3 5\, dx = $ _____

23) $\int_0^4 x\, dx = $ _____

24) $\int_0^\pi \cos(x)\, dx = $ _____

Chapter 6: Answers

1) $-\frac{1}{2x^2} + C$

2) $\frac{1}{3} \ln|5 + 3x| + C$

3) $\frac{e^{3x}(9x^2 - 6x + 2)}{27} + C$

4) $\frac{1}{3}\sqrt{(2x+1)^3} + C$

5) $\frac{2}{3} x^{\frac{3}{2}} + C$

6) $2\sqrt{x+4} + C$

7) $x \sin(x) + \cos(x) + C$

8) $x \ln(x) - x + C$

9) $\frac{8}{3}$

10) 2

13) $\frac{1}{8}$

14) $\frac{5}{2}$

17) 2

18) 1

21) 16

22) 15

11) 1

12) $\frac{\ln 2}{2}$

15) 3.39

16) -8

19) Divergent (does not converge)

20) Divergent (does not converge)

23) 8

24) 0

CHAPTER 7
Differential Equations

Math topics that you'll learn in this chapter:

- ☑ Introduction and applications
- ☑ Classification of Differential Equations
- ☑ First-Order Ordinary Differential Equations
- ☑ Linear Differential Equations
- ☑ Separable Differential Equations
- ☑ Slope Fields
- ☑ Euler's Method for Numerical Solutions
- ☑ Simple Growth and Decay
- ☑ Population Models

Introduction and applications

- Differential equations relate functions to their rates of change, using derivatives. They describe dynamic systems and phenomena.

- Solutions yield functions or set of functions that meet specified conditions, not numbers (or set of numbers) like we are used to in algebraic equations, for example.

- Differential equations can be classified as ordinary or partial, depending on whether they involve regular or partial derivatives. Their complexity ranges from simple linear forms to intricate nonlinear ones. Solutions can be exact or approximate.

- Differential equations originated in the 17th century with Newton and Leibniz's invention of calculus. Pioneers like Euler and Bernoulli expanded the field, applying them to physical phenomena, leading to vast applications in science, engineering, and beyond in subsequent centuries.

- Differential equations quantify how things change. For example:

  - **Population Growth**: A differential equation can describe how a population grows over time, factoring in birth and death rates.

  - **Fluid Dynamics**: They can describe how fluids move and behave, essential in aerodynamics or weather prediction.

  - **Electrical Circuits**: Differential equations represent voltages and currents in circuits, crucial for electronic design.

  - **Motion**: Newton's second law, $F = ma$, is a differential equation describing the relationship between an object's mass and its acceleration when a net force is applied.

- By solving these equations, we can predict future states, optimize systems, or control processes.

Classification of Differential Equations

- A **dependent variable** responds to changes in another variable. For example, in a study of sunlight's effect on plant growth, the plant's height (dependent) changes based on sunlight amount (independent).

- Differential equations can be classified based on several characteristics:

    1. **Order**: Refers to the highest derivative present. A differential equation with the highest derivative being the first derivative is called a first-order differential equation, and so on. For instance, $\frac{dy}{dx} + y = 0$ is first order, while $\frac{d^2y}{dx^2} + y = 0$ is second order.

    2. **Degree**: The degree is the power of the highest-order derivative. For example, the degree of $\left(\frac{d^2y}{dx^2}\right)^2 - 3\left(\frac{dy}{dx}\right)^4 + 2y = 0$ is 2, because the highest-order derivative in this equation is the derivative $\left(\frac{d^2y}{dx^2}\right)^2$. Typically, most differential equations encountered are of the first degree.

    3. **Linearity**: If the dependent variable and its derivatives appear linearly, i.e., aren't raised to any power or multiplied together, the differential equation is linear. Otherwise, it's nonlinear. For example, $\frac{dy}{dx} + y = 0$ and $\frac{dy}{dx} + x^2 = 0$ are linear, while $\frac{dy}{dx} + y^2 = 0$ and $y\frac{dy}{dx} = 0$ are nonlinear.

    4. **Homogeneity**: "a free term is a term that stands on its own, without being multiplied by the dependent variable or its derivatives". If the free term is zero, then the differential equation is homogeneous. If the free term exists (is non zero), the equation is non-homogeneous.
    $\frac{dy}{dx} = y$ is homogeneous, whereas $\frac{dy}{dx} = y + cos(x)$ is non-homogeneous.

    5. **Ordinary vs. Partial**: If the function depends on only one independent variable (like x), its differential equation involves ordinary derivatives and is termed an ordinary differential equation (ODE). If it involves derivatives with respect to multiple independent variables, and involves partial derivatives, it's a partial differential equation (PDE), for example $\frac{\partial u}{\partial x} + \frac{\partial u}{\partial y} = 0$. Nevertheless, partial differential equations are not typically included in high school mathematics.

First-Order Ordinary Differential Equations

- An ODE is a differential equation containing one or more unknown functions and their derivatives, but only with respect to one independent variable. The term "ordinary" here means that there are no partial derivatives involved.

- They take the general form $\frac{dy}{dx} = f(x, y)$ where f is a given function of two variables, x and y. The primary goal is to find the function y that satisfies this relation. $f\left(x, y, \frac{dy}{dx}\right) = 0$ is also used to denote the same concept.

- Here, the differential equation can involve x, y, and $\frac{dy}{dx}$ in any combination, and it isn't restricted to a linear format. So, this general form encompasses both linear and nonlinear first-order differential equations.

- There are various subclasses within first-order ODEs, such as separable, linear, and exact differential equations. Each type has its own unique methods for finding solutions.

Examples:

Example 1. Let's consider a simple first-order ODE: $\frac{dy}{dx} = y^2 + x$.

Solution: As you can see from y^2, it is not linear. You will learn later on that this kind of equation is called "separable" and how you can solve them.

Example 2. Consider the equation: $y \cos(x) = \frac{d(y^2)}{dx}$.

Solution: At first glance, it might not seem like a typical first-order ODE, but if you differentiate y^2 and use the chain rule, you'll see its first-order nature:

$\frac{d(y^2)}{dx} = 2y \frac{dy}{dx} \Rightarrow 2y \frac{dy}{dx} = y \cos(x)$. So: $\frac{dy}{dx} = \frac{1}{2} \cos(x)$.

The final form is a linear first-order ODE. The initial representation might have suggested nonlinearity, but if simplified, it's evident that the equation is linear.

Example 3. Here's a nonlinear first-order ODE: $y \frac{dy}{dx} = 1 + x^2 y$.

Solution: This differential equation contains products of y and its derivative (nonlinear), as well as a nonlinear term in $x^2 y$.

Linear Differential Equations

- Linear differential equations are a subclass of differential equations wherein the unknown function and its derivatives appear linearly, meaning they are not raised to any power higher than one and they are not multiplied together or contained within other functions.

- Essentially, these equations maintain a linear structure regarding the function and its derivatives.

  - **First-Order Linear ODEs**: A first-order linear ordinary differential equation (ODE) has the general form: $\frac{dy}{dx} + P(x) \cdot y = Q(x)$.

    Where $P(x)$ and $Q(x)$ are continuous functions of x in a certain interval.

    For example: $\frac{dy}{dx} + 3y = 5x$.

  - **Higher-Order Linear ODEs**: The general form of a linear ODE of order n is:

    $$f_n(x)y^{(n)} + f_{n-1}(x)y^{(n-1)} + \cdots + f_2(x)y'' + f_1(x)y' + f_0(x)y = g(x)$$

    Where $y^{(n)}$ represents the nth derivative of y with respect to x, and the functions $f_i(x)$ and $g(x)$ are given functions.

    For example, $y'' - y = e^x$.

  - **Homogeneity**: A linear differential equation is said to be homogeneous if the right side of the formula is zero (i.e., $g(x) = 0$ in the general form above). If $g(x)$ is not zero, then the equation is non-homogeneous.

  - **Solution Technique**: For simple differential equations, the "integrating factor method" is often used to find solutions. However, for more complex equations, methods such as "characteristic equations" and "undetermined coefficients" are employed to solve them.

  - **Linear PDEs (Partial Differential Equations)**: The concept of linearity also extends to partial differential equations. In linear PDEs, the dependent variable and its partial derivatives appear linearly.

Separable Differential Equations

- Separable differential equations are a specific class of ordinary differential equations (ODEs) that can be rewritten in a form that allows the variables to be separated, meaning all terms involving y (the dependent variable) can be written on one side and all terms involving x (the independent variable) on the other. This separation enables the equation to be integrated with respect to each variable independently.

- A separable differential equation can be shown in the form:
$$\frac{dy}{dx} = g(y) \cdot h(x)$$
Where $g(y)$ is a function solely of y and $h(x)$ is a function solely of x.

- To solve a separable differential equation:
    1. **Separate the Variables**: Rewrite the equation to have all terms involving y on one side and all terms involving x on the other. The goal is to get: $\frac{1}{g(y)} \cdot dy = h(x) \cdot dx$.
    2. **Integrate Both Sides**: Integrate each side of the equation with respect to its variable: $\int \frac{1}{g(y)} dy = \int h(x)\, dx$.
    3. **Solve for y (if possible)**: If the resulting equation can be solved explicitly for y, do so. Otherwise, the equation may serve as an implicit solution.

Example:

Consider the differential equation: $\frac{dy}{dx} = y \cdot x$.

Solution: Separate the variables: $\frac{dy}{y} = x \cdot dx$, if we Integrate both sides: $ln|y| + C_1 = \frac{x^2}{2} + C_2$, which can be simplified further using ln properties and the fact that subtraction of constants, will just yield a new constant:

$$ln|y| = \frac{x^2}{2} + C \Rightarrow e^{ln|y|} = e^{\frac{x^2}{2}+C} \Rightarrow |y| = e^{\frac{x^2}{2}+C}$$

This result provides a general solution to the differential equation in an **implicit** form. (The presence of the absolute value function means that y is not explicitly solved for in terms of x).

Slope Fields

- Slope fields, often referred to as direction fields, provide a visual means to understand the behavior of solutions to <u>first-order differential equations</u> without actually solving them.

- Essentially, they comprise a grid of short line segments, where each segment's slope corresponds to the value of the differential equation at that point.

- Each slope is calculated by trying out different numbers of x or y in the equation.

- By following these miniature slopes, one can sketch a curve representing the solution to the equation.

- The more slopes you draw (from the points closer to each other), the closer you get to the solution curve.

- This visualization prowess makes slope fields a cornerstone in the exploration of differential equations.

Example:

Consider the simple differential equation: $\frac{dy}{dx} = x + y$.

Solution: At the origin (0,0), the slope is 0, resulting in a horizontal line segment. However, at the point (1,1), the sum $x + y$ is equals 2, hence the line segment at that point has a slope of 2.

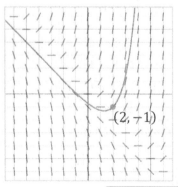

As one plots these segments across the plane, an intriguing pattern emerges, hinting at the solution curve's general behavior.

Euler's Method for Numerical Solutions

- Euler's method is a numerical approach used to approximate solutions to ordinary differential equations (ODEs). It provides a simple way to estimate the behavior of a system described by a differential equation.

- In this method, the idea is to break down the continuous differential equation into discrete steps. The derivative at a given point is approximated using the slope of the tangent line. By taking small steps in the independent variable using the current slope, the solution can be approximated incrementally. So, we assume no change in the slope over the interval, and the smaller the intervals are, the more accurate our approximation will become.

- There's a general formula for Euler's method. If you're given a first-order ODE of the form: $\frac{dy}{dx} = f(x,y)$ with an initial condition $y(x_0) = y_0$, then the recursive formula to approximate the solution using Euler's method is: $y_{n+1} = y_n + [h \times f(x_n, y_n)]$.

- Euler's method serves as a building block for more sophisticated numerical techniques and offers valuable insights into the behavior of differential equations.

Example:

Let's consider a first-order ODE: $\frac{dy}{dx} = x^2$, $y(0) = 1$.

Solution: Starting with an initial condition $y(0) = 1$, we can use Euler's method to estimate the value of y at different points. Assuming a step size of $h = 0.1$, the next approximation is given by:

$y(0.1) \approx y(0) + h \times f(0, y(0))$, where $f(x, y)$ represents the derivative (n^2 in this case). Using this formula, we can calculate: $y(0.1) \approx 1 + 0.1 \times (0)^2 = 1$. The process can be repeated to obtain approximate values of y at successive points, so:

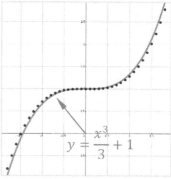

$y(0.2) \approx y(0.1) + h \times f(0.1, y(0.1)) = 1 + 0.1 \times (0.1^2) = 1.001$.

And so on: $y(0.3) = 1.005$, $y(0.4) = 1.014$ and $y(0.5) = 1.028$. These points represent the predicted y −values at corresponding x −values.

If we solve the equation by integrating both sides: $\frac{dy}{dx} = x^2 \Rightarrow dy = x^2\, dx$.

$\int dy = \int x^2\, dx \Rightarrow y = \frac{x^3}{3}$, and we can see that our approximation was accurate. (1 is added to the graph because of $y(0) = 1$)

Simple Growth and Decay

- Growth and decay equations Refer to equations modeling increasing or decreasing quantities. It involves a constant multiplying the current value.

  For example, $\frac{dy}{dx} = 0.1y$, models a 10% growth. If $y(0) = 100$, then $y(1) = 110$, because when integrating $\frac{dy}{dx} = 0.1y$, we get $y(x) = 100e^{0.1x}$, plugging in $x = 1$, we get 110.

- Simple growth and decay describe systems where the rate of change of a quantity is proportional to the current amount of that quantity. Mathematically, this relationship is given by $\frac{dy}{dt} = ky$, where y is the quantity of interest, t is time, and k is the proportionality constant. If $k > 0$, the system exhibits **exponential growth**, and if $k < 0$, it exhibits **exponential decay**. For example, the population of bacteria might grow exponentially in a nutrient-rich environment. Conversely, a radioactive substance might decay exponentially over time.

Example of Growth:

- Let's say bacteria population $P(t)$ doubles every hour. This can be modeled as $\frac{dP}{dt} = kP$ with $k = \ln 2$. If initially $P(0) = 100$, after one hour, $p(1) \approx 200$.

- Consider a city whose population increases by 5% each year due to migration and births. If the current population is 50,000, this can be modeled by the differential equation $\frac{dP}{dt} = 0.05P$.

  If $P(0) = 50000$, after one year, $P(1)$, would be 52,500.

Example of decay:

- Consider a radioactive substance with half-life of 3 hours. Its decay can be captured by $\frac{dN}{dt} = -kN$ with $k = \frac{\ln 2}{3}$, starting with $N(0) = 1000$ units, after three hours, we'd have $N(3) \approx 500$ units remaining.

- A cold beverage loses its fizz as time progresses. If the concentration of carbon dioxide decreases by 8%, every minute, this can be modeled by the differential equation $\frac{dC}{dt} = -0.08C$.

  If initially $C(0) = 100\%$, after one minute, $C(1) = 92\%$, and $C(2) = 92\% \times (1 - 0.08) = 84.64\%$.

Chapter 7: Differential Equations

Population Models

- Population models demonstrate how populations accelerate, reach a maximum growth rate, and stabilize as they approach the environment's carrying capacity. This model aids in understanding resource interactions and the influence of environmental factors on population dynamics.

- The predator-prey model, exemplified by the **Lotka-Volterra** equations, utilizes differential equations to elucidate the relationship between predator and prey populations.

- These models find practical applications in ecology, conservation biology, and resource management.

- One example of a population model is the exponential growth model. It can be represented by the differential equation: $\frac{dP}{dt} = rP$.

where P represents the population size, t represents time, and r is the growth rate.

This equation states that the rate of change of the population with respect to time is proportional to the population size.

Example: let's consider a population of bacteria with an initial population size of 100 bacteria ($P_0 = 100$) and a growth rate of 0.05 per hour ($r = 0.05$). Determine the population size after 3 hours.

Solution: We can use the exponential growth model to predict the population size at a given time.

The equation becomes: $\frac{dP}{dt} = 0.05P$, which is separable and can be rewritten as $\frac{dp}{p} = 0.05 \, dt$, and by integrating both sides: $\int \frac{dP}{p} = \int 0.05 \, dt$.

Then, $ln|p| + C_1 = 0.05t + C_2$. So: $|p| = e^{0.05t+C}$, and we know that population can't be negative, so: $p = e^{0.05t+C}$.

Substituting the values, we have: $P(t) = 100 \times e^{0.05t}$.

$P(3) = 100 \times e^{0.05 \times 3} = 100 \times e^{0.15} \approx 100 \times 1.1618 \approx 116.18$.

According to the model, the population of bacteria would be approximately 116.18 after 3 hours of exponential growth.

Chapter 7: Practices

✏ **Solve the differential equations.**

1) $\frac{dy}{dx} = 2y$

2) $\frac{dy}{dx} + 3y = 0$

3) $\frac{dy}{dx} = 4x$

4) $\frac{dy}{dx} = y\,tan(x)$

✏ **Sketch the slope field for the differential equation.**

5) $\frac{dy}{dx} = 5y + 1$

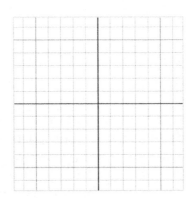

✏ **Model the given phenomena.**

6) An initially small amount of substance grows at a rate proportional to its current quantity. Write a differential equation to model this. _____

✏ **Solve the separable differential equation.**

7) $\frac{dy}{dx} = xy^2$

8) $\frac{dy}{dx} = -2xy$

Solve the Growth and Decay problems.

9) A certain species of plant grows continuously at a rate proportional to its current size. If the plant is initially 10 cm tall and grows to 15 cm tall in one week, how tall will it be at the end of 3 weeks?

10) A video on social media is becoming viral. If it gains views at a rate proportional to its current views and it initially has 1,000 views, but then reaches 3,000 views in 2 days, how many views will it have after 5 days?

11) A certain medicine in the body decays continuously. If initially there's 100 mg, and after 4 hours only 60 mg remain, how much will be left after 12 hours?

12) A gadget's battery drains at a rate proportional to its current charge. If it starts at 100% and drops to 75% in 3 hours, what will the charge be after 9 hours?

Find the degree of given differential equations.

13) $\frac{dy}{dx} + y^2 = 1$

14) $\frac{d^3y}{dx^3} + \left(\frac{d^2y}{dx^2}\right)^4 = 0$

15) $y^4 \cdot (y')^3 + (y'')^5 - \sin(y) = x^2$

16) $(y'')^2 - 3(y')^3 + 2y = 0$

Find the order of given differential equations.

17) $4\frac{dy}{dx} + y^3 = 0$

18) $\left(\frac{d^2y}{dx^2}\right)^3 - 2y^3 + x^4 = 0$

19) $\frac{d^3y}{dx^3} + 2\frac{dy}{dx} - y = \sin(x)$

20) $y'' + (y')^2 - 3y = e^x$

Chapter 7: Answers

1) $y(x) = Ce^{2x}$

2) $y(x) = Ce^{-3x}$

3) $y(x) = 2x^2 + C$

4) $y(x) = C\sec(x)$

5) $y(x) = \dfrac{e^{5x+C'} - 1}{5}$

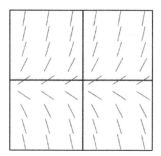

6) $\dfrac{dy}{dt} = ky$

7) $y(x) = -\dfrac{2}{x^2 + C}$ 8) $y(x) = Ce^{-x^2}$

9) $33.75\ cm$ 11) $\approx 21.6\ mg$

10) $\approx 15{,}588$ views   12) $\approx 42.19\ \%$

13) 1   15) 5

14) 1   16) 2

17) First order   19) Third order

18) Second order   20) Second order

CHAPTER 8
Analytic Geometry

Math topics that you'll learn in this chapter:

- ☑ Ellipses, Parabolas and Hyperbolas
- ☑ Polar Coordinates
- ☑ Converting between polar coordinates and rectangular coordinates
- ☑ Graphing polar equations
- ☑ Applications of polar coordinates

Ellipses, Parabolas, and Hyperbolas

- Ellipses, parabolas, and hyperbolas are all types of conic sections.

- An **ellipse** is a curve formed by intersecting a cone with a plane at an angle, resulting in a closed curve with two foci points. It can be described using the equation $\left(\frac{x}{a}\right)^2 + \left(\frac{y}{b}\right)^2 = 1$, where a and b represent the lengths of the semi-major and semi-minor axes. For example, you can see the ellipse with the formula $\frac{x^2}{4} + \frac{y^2}{9} = 1$.

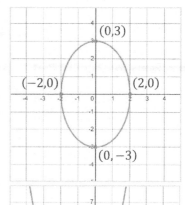

- A **circle** is a special form of an ellipse where both foci are at the center.

- A **parabola** is a curve obtained by slicing a cone parallel to its side. It has a single focus point and a directrix, and its equation can be expressed as $y = ax^2 + bx + c$, where a, b, and c are coefficients determining its shape and position. For example, you can see the parabola with the formula $y = x^2 + 2x + 1$.

- A **hyperbola** is formed by slicing a cone at an angle greater than that of an ellipse. It has two branches that are symmetrical to each other, and its equation can be represented as $\left(\frac{x}{a}\right)^2 - \left(\frac{y}{b}\right)^2 = 1$ or $\left(\frac{y}{b}\right)^2 - \left(\frac{x}{a}\right)^2 = 1$, where a and b are constants determining the shape and position. You can see the hyperbola with the following formula: $\left(\frac{y}{2}\right)^2 - \left(\frac{x}{1}\right)^2 = 1 \Rightarrow \frac{y^2}{4} - \frac{x^2}{1} = 1$.

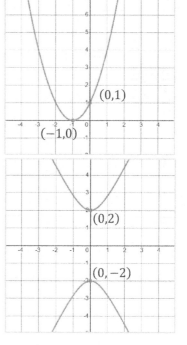

  Circle   Ellipse   Parabol   Hyperbola

Polar Coordinates

- Polar coordinates are a two-dimensional coordinate system, just like the Cartesian (or rectangular) coordinate system. However, instead of using horizontal and vertical displacements (x and y) to locate points, polar coordinates locate points in the plane using a distance and an angle.

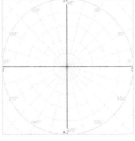

- The polar coordinate system is particularly useful in situations where the problem has rotational symmetry, for example in the study of circular and elliptical paths in physics and engineering.
- Here's a breakdown of polar coordinates:
  - **Radius (r):** The first element of a polar coordinate is r, which represents the direct distance from the origin (O) to the point in the plane. The value of r can be any non-negative real number.
  - **Theta (θ):** The second element of a polar coordinate is θ, an angle measured counterclockwise from the positive x-axis to the line segment that joins the point to the origin.
- For example, the polar coordinates $(r, \theta) = \left(3, \frac{\pi}{2}\right)$ represent the point that is 3 units away from the origin, in the direction $\frac{\pi}{2}$ radians (or 90 degrees) counterclockwise from the positive x-axis. This point would be at $(0,3)$ in Cartesian coordinates.

Example:

Find the angle, in degrees, between the positive x-axis and the line segment connecting the origin to the point $(3,4)$ in the polar coordinate system.

Solution: To find the angle between the positive x-axis and the line segment connecting the origin to the point $(3,4)$ in the polar coordinate system, we can use the angular coordinate (θ) of the point. The angular coordinate (θ) represents the angle measured counterclockwise from the positive x-axis to the line segment connecting the origin to the point. In this case, θ can be found using the tangent function:

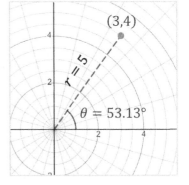

$$\theta = \arctan\left(\frac{y}{x}\right) = \arctan\left(\frac{4}{3}\right)$$

Using a calculator, we can find that $\arctan\left(\frac{4}{3}\right)$ is approximately $53.13°$.

Therefore, the angle between the positive x-axis and the line segment connecting the origin to the point $(3,4)$ is approximately $53.13°$.

Converting Between Polar and Rectangular Coordinates

Rectangular to Polar Conversion:
- Given a point in rectangular coordinates (x, y), we can find the polar coordinates (r, θ) using the following equations:
  - The radial coordinate 'r' is found using the Pythagorean theorem:
  $$r = \sqrt{(x^2 + y^2)}$$
  - The angular coordinate 'θ' can be found using trigonometric functions. The tangent of 'θ' is $\frac{y}{x}$, so $\theta = arctan\left(\frac{y}{x}\right)$. However, the value of 'θ' must be adjusted based on the quadrant of the point:
    - Quadrant I: $\theta = arctan\left(\frac{y}{x}\right)$
    - Quadrant II: $\theta = \pi + arctan\left(\frac{y}{x}\right)$
    - Quadrant III: $\theta = \pi + arctan\left(\frac{y}{x}\right)$
    - Quadrant IV: $\theta = 2\pi + arctan\left(\frac{y}{x}\right)$

Polar to Rectangular Conversion:
- Given a point in polar coordinates (r, θ), we can find the rectangular coordinates (x, y) using the following equations:
  - The x–coordinate can be found using the formula: $x = r \times cos(\theta)$.
  - The y–coordinate can be found using the formula: $y = r \times sin(\theta)$.
- Note: Make sure that the angle θ is in the correct form (usually radians) for the trigonometric functions.
- Remember, these conversion methods are applicable in a 2D plane. If you are working with 3D space, you would use cylindrical or spherical coordinates instead of polar.

Example:

Sarah is located at coordinates $(1, \sqrt{3})$ in the rectangular coordinate system. Convert Sarah's coordinates to polar coordinates.

Solution: To convert Sarah's coordinates $(1, \sqrt{3})$ from rectangular to polar coordinates, we can use the following formulas: $r = \sqrt{x^2 + y^2}$, $\theta = arctan\left(\frac{y}{x}\right)$. Plugging in the values for Sarah's coordinates, we have:

$r = \sqrt{1^2 + \left(\sqrt{3}\right)^2} = \sqrt{1 + 3} = \sqrt{4} = 2$, and $\theta = arctan\left(\frac{\sqrt{3}}{1}\right) = arctan(\sqrt{3}) = \frac{\pi}{3}$.

Therefore, Sarah's coordinates in the polar form are approximately $\left(2, \frac{\pi}{3}\right)$.

bit.ly/3JBukgU

Graphing Polar Equations

- Graphing polar equations involves plotting points that satisfy the equation on a polar coordinate plane, which uses radial and angular coordinates instead of the conventional Cartesian coordinates.

- Here's a brief step-by-step process:

  - **Understand the Polar Coordinate System**: This system is defined by a distance (r) from a central point (the origin) and an angle (θ) from the positive x-axis.

  - **Identify the Polar Equation**: Polar equations can take various forms, such as circles, spirals, or roses.

  - **Create a Table of Values**: Substitute various values of θ into the equation and solve for r to understand what points satisfy the equation.

  - **Plot Points**: Plot these (r, θ) values on your polar graph. The angle θ is counterclockwise from the x-axis, and r is the distance from the origin.

  - **Connect the Dots**: Draw a smooth curve that connects these plotted points to form the graph of the polar equation.

  - **Analyze the Graph**: Identify the graph's properties, such as symmetry or intercepts, and determine the *maximum* and *minimum* values.

Example:

Consider the polar equation $r = 4 + 2\cos(\theta)$. Sketch the graph of the polar equation on the coordinate plane.

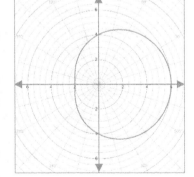

Solution: To sketch the graph, we can plot points for different values of θ and corresponding values of r: When $\theta = 0°$, $r = 6$. Plot a point at $(6,0)$.

When $\theta = 45°$, $r = 4 + 2\cos(45°) = 4 + 2\left(\frac{\sqrt{2}}{2}\right) = 4 + \sqrt{2} \approx 5.41$. Plot a point at $(5.41, 45°)$. When $\theta = 90°$, $r = 4$. Plot a point at $(4, 90°)$. When $\theta = 135°$, $r = 4 + 2\cos(135°) = 4 + 2\left(-\frac{\sqrt{2}}{2}\right) = 4 - \sqrt{2} \approx 2.59$. Plot a point at $(2.59, 135°)$. When $\theta = 180°$, $r = 2$. Plot a point at $(2, 180°)$. Continue this process until you complete the full $360°$ rotation. Once all the points are plotted, connect them smoothly to form the graph of the polar equation $r = 4 + 2\cos(\theta)$.

Applications of Polar Coordinates

- Polar coordinates are used extensively in various fields of mathematics, physics, engineering, computer science, and more. Here are some specific applications:

  - **Physics and Engineering**: Polar coordinates are often used in physics and engineering to describe phenomena that have a clear point of origin or center. For example, in mechanics, polar coordinates are used to analyze circular motion, oscillatory motion, and wave phenomena. In electrical engineering, they are used in signal processing and system analysis (especially in the frequency domain using the Fourier Transform).

  - **Astronomy**: Astronomers use polar coordinates to describe the location of stars and other celestial bodies. By using the observer as the origin point, the angle from a reference direction (like due north) and the distance to the star can be used to pinpoint its location.

  - **Computer Graphics and Game Design**: Polar coordinates are used in computer graphics and game design, especially for objects that move or exist in a circular pattern. For example, polar coordinates can be used to rotate an object around a point or to create circular paths for objects to follow.

  - **Robotics**: In robotics, polar coordinates are used in the design of robotic arms and other machinery. The polar coordinate system makes it easier to calculate and control the movement of these machines.

  - **Geographic Information Systems (GIS)**: Polar coordinates are used in geographic information systems to describe locations on the earth's surface. This is especially useful in navigation and cartography.

  - **Mathematics**: In calculus, polar coordinates can be used to solve certain types of integrals and differential equations more easily than with Cartesian coordinates. Also, in complex analysis, complex numbers are often represented in a polar form.

  - **Meteorology**: Polar coordinates are used in meteorology to represent wind direction and speed. The angle represents the wind direction, and the radial distance represents the speed. These are just a few examples of how polar coordinates are applied in various fields. In general, any situation where a phenomenon is naturally centered around a point or naturally moves in a circular or spherical pattern is a candidate for the use of polar coordinates.

Chapter 8: Practices

✏️ **Solve.**

1) Find the distance between the origin and the point $\left(5, \frac{\pi}{3}\right)$ in the polar coordinate system.

2) A satellite is orbiting the Earth in a polar orbit with a radius of 1000 kilometers. Determine the polar coordinates of the satellite's position after it completes a full orbit around the Earth.

3) A radar station is tracking an aircraft that is flying at a distance of 500 meters from the origin at an angle of 30 degrees counterclockwise from the positive x–axis. Determine the polar coordinates of the aircraft's position.

4) A target is positioned at a distance of 10 units from the origin at an angle of 60 degrees counterclockwise from the positive x–axis. What are the polar coordinates of the target's position?

✏️ **Determine the rectangular coordinates of points.**

5) $(6, 45°) = $ _____

6) $(8, 30°) = $ _____

7) $\left(12, \frac{2\pi}{3}\right) = $ _____

8) $\left(1, \frac{\pi}{2}\right) = $ _____

9) $(5, 60°) = $ _____

10) $\left(8, \frac{\pi}{4}\right) = $ _____

✏️ **Sketch the graph of the polar equations on the coordinate plane.**

11) $r = 6 - 6\cos(\theta)$

12) $r = 9 + 5\cos(\theta)$

Effortless Math Education

Chapter 8: Answers

1) 5 unit

2) (1000,0°)

5) (4.24,4.24)

6) (6.93,4)

7) (−6,10.39)

3) (500,30°)

4) (10,60°)

8) (0,1)

9) (2.5,4.33)

10) (5.66,5.66)

11)

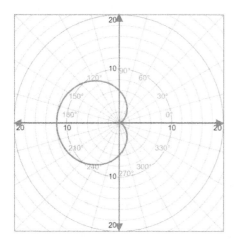

12)

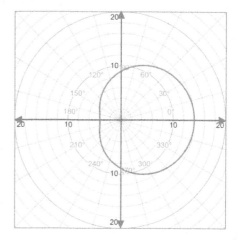

CHAPTER
9 Complex Numbers

Math topics that you'll learn in this chapter:

☑ Complex Numbers Addition and Subtraction
☑ Multiplying and Dividing Complex Numbers
☑ Rationalizing Imaginary Denominators

Complex Numbers Addition and Subtraction

- A complex number is expressed in the form $a + bi$, where a and b are real numbers, and i, is called an imaginary number and is a solution of the equation $i^2 = -1$, Which is normally considered undefined.

- Complex Numbers solutions to equations with no real solutions and are fundamental in advanced mathematics, physics, and engineering for modeling periodic and oscillatory behavior.

- They enable calculations beyond real number limitations, crucial for quantum mechanics, signal processing, and solving otherwise intractable differential equations.

- For adding or subtracting complex numbers, we need to add/subtract real numbers together, and imaginary parts together:

$$(a + bi) \pm (c + di) = (a \pm c) + (b \pm d)i$$

Examples:

Example 1. Solve: $(8 + 4i) + (6 - 2i)$.

Solution: Combine like terms: $8 + 4i + 6 - 2i = 14 + 2i$.

Example 2. Solve: $(10 + 8i) + (8 - 3i)$.

Solution: Group like terms: $10 + 8i + 8 - 3i = 18 + 5i$.

Example 3. Solve: $(-5 - 3i) - (2 + 4i)$.

Solution: Remove parentheses by multiplying -1 to the second parentheses:

$(-5 - 3i) - (2 + 4i) = -5 - 3i - 2 - 4i$.

Combine like terms: $-5 - 3i - 2 - 4i = -7 - 7i$.

Example 4. Solve: $(3\pi - \pi i) - (\pi + 3\pi i)$.

Solution: $3\pi - \pi i - \pi - 3\pi i = 2\pi - 4\pi i$.

Multiplying and Dividing Complex Numbers

- You can use the following rule to multiply imaginary numbers. Remember that: $i^2 = -1$: $(a + bi) \cdot (c + di) = (ac - bd) + (ad + bc)i$.

  You can reach this by multiplying the components like you would normally do when multiplying expressions, meaning you have to multiply a by both c and di, and so on:

  $$ac + adi + bci + bdi^2 = ac + adi + bci + (bd \times (-1))$$

  $$= (ac - bd) + (ad + bc)i$$

- To divide complex numbers, you need to find the conjugate of the denominator. The conjugate of a complex number is formed by changing the sign of the imaginary part: The conjugate of $(a + bi)$ is $(a - bi)$, and vice versa.

- Dividing complex numbers: $\frac{a+bi}{c+di} = \frac{a+bi}{c+di} \times \frac{c-di}{c-di} = \frac{ac+bd}{c^2+d^2} + \frac{bc-ad}{c^2+d^2}i$.

Examples:

Example 1. Solve: $\frac{6-2i}{2+i}$.

Solution: The conjugate of $(2 + i)$ is $(2 - i)$. Use the rule for dividing complex numbers:

$\frac{a+bi}{c+di} = \frac{a+bi}{c+di} \times \frac{c-di}{c-di} = \frac{ac+bd}{c^2+d^2} + \frac{bc-ad}{c^2+d^2}i$.

Therefore: $\frac{6-2i}{2+i} \times \frac{2-i}{2-i} = \frac{6\times(2)+(-2)(1)}{2^2+(1)^2} + \frac{-2\times2-(6)(1)}{2^2+(1)^2}i = \frac{10}{5} + \frac{-10}{5}i = 2 - 2i$.

Example 2. Solve: $(2 - 3i)(6 - 3i)$.

Solution: $(2 \times 6) + (2 \times -3i) + (-3i \times 6) + (-3i)^2 = 12 - 6i - 18i - 9 = 3 - 24i$.

Example 3. Solve: $\frac{3-2i}{4+i}$.

Solution: Use the rule for dividing complex numbers:

Therefore: $\frac{3-2i}{4+i} \times \frac{4-i}{4-i} = \frac{(3\times4+(-2i)\times(-i))+(-2\times4-3\times1)i}{4^2-i^2} = \frac{10-11i}{17} = \frac{10}{17} - \frac{11}{17}i$.

Rationalizing Imaginary Denominators

- Rationalizing imaginary denominators involves multiplying numerator and denominator by the conjugate of the denominator to remove the imaginary number from the denominator, creating a real number denominator.

- Rationalizing imaginary denominators makes complex fractions clearer and comparisons easier by ensuring denominators are real, simplifying calculations, and aligning with standard forms, especially useful historically and in practical applications like engineering where real numbers are often expected.

Examples:

Example 1. Solve: $\frac{4-3i}{6i}$.

Solution: Multiply both numerator and denominator by $\frac{i}{i}$:

$\frac{4-3i}{6i} = \frac{4-3i}{6i} \times \frac{i}{i}$. Therefore:

$\frac{4-3i}{6i} = \frac{(4-3i)(i)}{6i(i)} = \frac{(4)(i)-(3i)(i)}{6(i^2)} = \frac{4i-3i^2}{6(-1)} = \frac{4i-3(-1)}{-6} = \frac{4i}{-6} + \frac{3}{-6} = -\frac{1}{2} - \frac{2}{3}i$.

Example 2. Solve: $\frac{6i}{2-i}$.

Solution: Multiply both numerator and denominator by the conjugate $\frac{2+i}{2+i}$:

$\frac{6i}{2-i} = \frac{6i(2+i)}{(2-i)(2+i)}$. Apply complex arithmetic rule: $(a+bi)(a-bi) = a^2 + b^2$.

Therefore: $2^2 + (-1)^2 = 5$, then: $\frac{6i(2+i)}{(2-i)(2+i)} = \frac{-6+12i}{5} = -\frac{6}{5} + \frac{12}{5}i$.

Example 3. Solve: $\frac{8-2i}{2i}$.

Solution: Factor 2 from both sides: $\frac{8-2i}{2i} = \frac{2(4-i)}{2i}$, divide both sides by 2:

$\frac{2(4-i)}{2i} = \frac{(4-i)}{i}$. Multiply both numerator and denominator by $\frac{i}{i}$:

$\frac{(4-i)}{i} = \frac{(4-i)}{i} \times \frac{i}{i} = \frac{(4i-i^2)}{i^2} = \frac{1+4i}{-1} = -1 - 4i$.

Chapter 9: Practices

✎ **Simplify.**

1) $(-4i) - (7 - 2i) =$ _____

2) $(-3 - 2i) - (2i) =$ _____

3) $(8 - 6i) + (-5i) =$ _____

4) $(-3 + 6i) - (-9 - i) =$ _____

5) $(-5 + 15i) - (-3 + 3i) =$ _____

6) $(-14 + i) - (-12 - 11i) =$ _____

7) $(-18 - 3i) + (11 + 5i) =$ _____

8) $(-11 - 9i) - (-9 - 3i) =$ _____

9) $-8 + (2i) + (-8 + 6i) =$ _____

10) $(-2 - i)(4 + i) =$ _____

11) $(2 - 2i)^2 =$ _____

12) $(4 - 3i)(6 - 6i) =$ _____

13) $(5 + 4i)^2 =$ _____

14) $(4i)(-i)(2 - 5i) =$ _____

15) $(2 - 8i)(3 - 5i) =$ _____

16) $\frac{9i}{3-i} =$ _____

17) $\frac{2+4i}{14+4i} =$ _____

18) $\frac{5+6i}{-1+8i} =$ _____

19) $\frac{-8-i}{-4-6i} =$ _____

20) $\frac{-1+5i}{-8-7i} =$ _____

21) $\frac{-2-9i}{-2+7i} =$ _____

22) $\frac{-8}{-5i} =$ _____

23) $\frac{-5}{-i} =$ _____

24) $\frac{3}{5i} =$ _____

25) $\frac{6}{-4i} =$ _____

26) $\frac{-6-i}{-1+6i} =$ _____

27) $\frac{-9-3i}{-3+3i} =$ _____

28) $\frac{4i+1}{-1+3i} =$ _____

29) $\frac{6-3i}{2-i} =$ _____

30) $\frac{-5+2i}{2-3i} =$ _____

Chapter 9: Answers

1) $-7 - 2i$

2) $-3 - 4i$

3) $8 - 11i$

4) $6 + 7i$

5) $-2 + 12i$

6) $-2 + 12i$

7) $-7 + 2i$

8) $-2 - 6i$

9) $-16 + 8i$

10) $-7 - 6i$

11) $-8i$

12) $6 - 42i$

13) $9 + 40i$

14) $8 - 20i$

15) $-34 - 34i$

16) $-\frac{9}{10} + \frac{27}{10}i$

17) $\frac{11}{53} + \frac{12}{53}i$

18) $\frac{43}{65} - \frac{46}{65}i$

19) $\frac{19}{26} - \frac{11}{13}i$

20) $-\frac{27}{113} - \frac{47}{113}i$

21) $-\frac{59}{53} + \frac{32}{53}i$

22) $\frac{-8}{5}i$

23) $-5i$

24) $-\frac{3}{5}i$

25) $\frac{3}{2}i$

26) i

27) $1 + 2i$

28) $\frac{11}{10} - \frac{7}{10}i$

29) 3

30) $-\frac{16}{13} - \frac{11}{13}i$

Time to Test

Time to refine your Math skill with a practice test.

In this section, there are two complete Calculus Tests. Take these tests to simulate the test day experience. After you've finished, score your test using the answer keys.

Before You Start

- You'll need a pencil and a calculator to take the test.
- For each question, there are four possible answers. Choose which one is best.
- It's okay to guess. There is no penalty for wrong answers.
- Use the answer sheet provided to record your answers.
- **Calculator is permitted for Calculus Test.**
- After you've finished the test, review the answer key to see where you went wrong.

Good luck!

Calculus

Practice Test 1

2024

Total number of questions: 60

Time: <u>No time limit</u>

Calculator is permitted for Calculus Test.

Calculus Practice Test 1 Answer Sheet

Remove (or photocopy) this answer sheet and use it to complete the practice test.

Calculus Practice Test 1 Answer Sheet

#		#		#	
1	Ⓐ Ⓑ Ⓒ Ⓓ	21	Ⓐ Ⓑ Ⓒ Ⓓ	41	Ⓐ Ⓑ Ⓒ Ⓓ
2	Ⓐ Ⓑ Ⓒ Ⓓ	22	Ⓐ Ⓑ Ⓒ Ⓓ	42	Ⓐ Ⓑ Ⓒ Ⓓ
3	Ⓐ Ⓑ Ⓒ Ⓓ	23	Ⓐ Ⓑ Ⓒ Ⓓ	43	Ⓐ Ⓑ Ⓒ Ⓓ
4	Ⓐ Ⓑ Ⓒ Ⓓ	24	Ⓐ Ⓑ Ⓒ Ⓓ	44	Ⓐ Ⓑ Ⓒ Ⓓ
5	Ⓐ Ⓑ Ⓒ Ⓓ	25	Ⓐ Ⓑ Ⓒ Ⓓ	45	Ⓐ Ⓑ Ⓒ Ⓓ
6	Ⓐ Ⓑ Ⓒ Ⓓ	26	Ⓐ Ⓑ Ⓒ Ⓓ	46	Ⓐ Ⓑ Ⓒ Ⓓ
7	Ⓐ Ⓑ Ⓒ Ⓓ	27	Ⓐ Ⓑ Ⓒ Ⓓ	47	Ⓐ Ⓑ Ⓒ Ⓓ
8	Ⓐ Ⓑ Ⓒ Ⓓ	28	Ⓐ Ⓑ Ⓒ Ⓓ	48	Ⓐ Ⓑ Ⓒ Ⓓ
9	Ⓐ Ⓑ Ⓒ Ⓓ	29	Ⓐ Ⓑ Ⓒ Ⓓ	49	Ⓐ Ⓑ Ⓒ Ⓓ
10	Ⓐ Ⓑ Ⓒ Ⓓ	30	Ⓐ Ⓑ Ⓒ Ⓓ	50	Ⓐ Ⓑ Ⓒ Ⓓ
11	Ⓐ Ⓑ Ⓒ Ⓓ	31	Ⓐ Ⓑ Ⓒ Ⓓ	51	Ⓐ Ⓑ Ⓒ Ⓓ
12	Ⓐ Ⓑ Ⓒ Ⓓ	32	Ⓐ Ⓑ Ⓒ Ⓓ	52	Ⓐ Ⓑ Ⓒ Ⓓ
13	Ⓐ Ⓑ Ⓒ Ⓓ	33	Ⓐ Ⓑ Ⓒ Ⓓ	53	Ⓐ Ⓑ Ⓒ Ⓓ
14	Ⓐ Ⓑ Ⓒ Ⓓ	34	Ⓐ Ⓑ Ⓒ Ⓓ	54	Ⓐ Ⓑ Ⓒ Ⓓ
15	Ⓐ Ⓑ Ⓒ Ⓓ	35	Ⓐ Ⓑ Ⓒ Ⓓ	55	Ⓐ Ⓑ Ⓒ Ⓓ
16	Ⓐ Ⓑ Ⓒ Ⓓ	36	Ⓐ Ⓑ Ⓒ Ⓓ	56	Ⓐ Ⓑ Ⓒ Ⓓ
17	Ⓐ Ⓑ Ⓒ Ⓓ	37	Ⓐ Ⓑ Ⓒ Ⓓ	57	Ⓐ Ⓑ Ⓒ Ⓓ
18	Ⓐ Ⓑ Ⓒ Ⓓ	38	Ⓐ Ⓑ Ⓒ Ⓓ	58	Ⓐ Ⓑ Ⓒ Ⓓ
19	Ⓐ Ⓑ Ⓒ Ⓓ	39	Ⓐ Ⓑ Ⓒ Ⓓ	59	Ⓐ Ⓑ Ⓒ Ⓓ
20	Ⓐ Ⓑ Ⓒ Ⓓ	40	Ⓐ Ⓑ Ⓒ Ⓓ	60	Ⓐ Ⓑ Ⓒ Ⓓ

Calculus Practice Test 1

1) Which of the following is the domain of the function $f(x) = \sqrt{x-5}$?

   A. $x > 5$

   B. $x \leq 5$

   C. $x \geq 5$

   D. $\mathbb{R}$

2) The function $g(x) = |x|$ is not differentiable at:

   A. $x = 1$

   B. $x = 0$

   C. $x = -1$

   D. Any x value

3) Given $f(x) = 2x + 1$ and $g(x) = x^2$, what is $f(g(3))$?

   A. 20

   B. 19

   C. 10

   D. 7

4) If $f^{-1}(x)$ is the inverse of $f(x)$, then $f(f^{-1}(a)) =$:

   A. f(a)

   B. a

   C. 1

   D. 0

5) If $f(x) = x^2$, the inverse function $f^{-1}(x)$ is:

   A. x^2

   B. $\sqrt{x}$

   C. $-\sqrt{x}$

   D. $-x$

6) If $f(x) = x^2$, then the graph of $f(x-3) + 1$ results from:

   A. Shifting $f(x)$ three units right and one unit up

   B. Shifting $f(x)$ three units left and one unit up

   C. Shifting $f(x)$ three units right and one unit down

   D. Shifting $f(x)$ three units left and one unit down

7) The function $f(x) = -x^3 + 1$, reflects $y = x^3 + 1$ over the:

   A. x−axis

   B. y−axis

   C. Line $y = x$

   D. Origin

8) Which of the following sequences is arithmetic?

   A. 3, 6, 12, 24, ⋯

   B. 2, 5, 9, 13, ⋯

   C. 1, 4, 9, 16, ⋯

   D. 5, 3, 1, −1, ⋯

9) The nth term of an arithmetic sequence is given by $a_n = 2n + 3$. What is the 5th term?

   A. 15

   B. 13

   C. 10

   D. 5

10) Which of the following sequences is geometric?

    A. 1, 3, 5, 7, ⋯

    B. 4, 2, 1, 0.5, ⋯

    C. 1, 1, 2, 2, ⋯

    D. 2, 6, 12, 20, ⋯

11) What is the sum of the first 6 terms of the arithmetic series with a common difference of 4 and a first term of 3?

    A. 107

    B. 97

    C. 92

    D. 78

12) For a geometric series, if the first term is 3 and the common ratio is 2, what is the sum of the first 3 terms?

    A. 21

    B. 15

    C. 12

    D. 9

13) If the sum of an infinite geometric series is 8 and the first term is 4, the common ratio is:

   A. 0.25

   B. 0.5

   C. 2

   D. 4

14) Which of the following series is convergent?

   A. $1 + 2 + 3 + 4 + \cdots$

   B. $1 - \frac{1}{2} + \frac{1}{4} - \frac{1}{8} + \cdots$

   C. $1 + 1 + 1 + 1 + \cdots$

   D. $2 + 4 + 6 + 8 + \cdots$

15) The limit of $f(x) = x^2$ as x approaches 2 is:

   A. 0

   B. 2

   C. 4

   D. None

16) What is $\lim_{x \to 3}(2x - 5)$?

   A. 0

   B. 1

   C. 6

   D. 10

17) The function $f(x) = \frac{x^2-9}{x-3}$ is:

    A. Continuous at $x = 3$

    B. Discontinuous at $x = 3$

    C. Defined at $x = 3$

    D. All of the above

18) Which of the following is true about $\lim\limits_{x \to 0} \frac{sin(x)}{x}$?

    A. The limit is 0

    B. The limit is 1

    C. The limit does not exist.

    D. None of the above

19) If $f(x) = x^3 - x$, the limit as x approaches 1 is:

    A. 0

    B. 1

    C. 2

    D. None

20) The function $f(x) = \sqrt{x}$ is continuous over:

    A. All real numbers

    B. $x \geq 0$

    C. $x > 0$

    D. None of the above

21) $\lim_{x \to 0} x^2 \sin\left(\frac{1}{x}\right)$ is:

   A. 1

   B. 0

   C. −1

   D. Undefined

22) What is $\lim_{x \to 2^-} \frac{1}{x-2}$?

   A. 0

   B. Undefined

   C. ∞

   D. −∞

23) If $f(x) = \frac{1}{x}$, the limit as x approaches 0 from the right is:

   A. 0

   B. 1

   C. ∞

   D. Undefined

24) The function $f(x) = \frac{x}{|x|}$ is discontinuous at:

   A. $x = 0$

   B. $x = 1$

   C. $x = -1$

   D. $x = 2$

25) A function is said to be continuous at a point if:

   A. The function is defined at that point.

   B. The limit of the function exists at that point.

   C. The function's value equals its limit at that point.

   D. None

26) For which value of c is $f(x) = x^2 + cx + 1$ continuous everywhere?

   A. 0

   B. 1

   C. 2

   D. $\mathbb{R}$

27) $\lim\limits_{x \to 1} \dfrac{x^2-1}{x-1}$ equals:

   A. 0

   B. 1

   C. 2

   D. Undefined

28) The point of discontinuity for the function $f(x) = \dfrac{x^2-4}{x-2}$ is:

   A. 4

   B. 2

   C. 0

   D. -2

29) What is $\lim_{x \to 0^+} x \ln(x)$?

   A. -1

   B. 0

   C. 1

   D. Undefined

30) The limit of $\tan(x)$ as x approaches $\frac{\pi^-}{2}$ is:

   A. $+\infty$

   B. 1

   C. 0

   D. $-\infty$

31) Using the product rule, find the derivative of $f(x) = x^2 \cdot e^x$.

   A. $x^2 + 2xe^x$

   B. $x^2 e^x + 2xe^x$

   C. $2x + e^x$

   D. $2xe^x$

32) The chain rule is used to find the derivative of:

   A. $f(g(x))$

   B. $f(x) - g(x)$

   C. $f(x) \cdot g(x)$

   D. $\frac{f(x)}{g(x)}$

33) The derivative of $f(x) = x^2 - 4x + 7$ is:

   A. $2x - 4$

   B. x^2

   C. $x - 4$

   D. $2x + 7$

34) If $y = \ln(5x)$, then y' is:

   A. 5

   B. $\frac{5}{x}$

   C. $\frac{1}{x}$

   D. $\frac{1}{5x}$

35) The derivative of $f(x) = arcsin(x)$ is:

   A. $\frac{1}{\sqrt{1-x^2}}$

   B. $\sqrt{1-x^2}$

   C. $\frac{1}{x\sqrt{1-x^2}}$

   D. $\frac{x}{\sqrt{1-x^2}}$

36) The derivative of the function $y = 5e^{2x}$ is:

   A. $5e^{2x}$

   B. $10e^{2x}$

   C. $2e^{2x}$

   D. $10e^x$

37) If $h(x) = sec(x)$, what is $h'(x)$?

   A. $sec(x) \, tan(x)$

   B. $csc(x) \, cot(x)$

   C. $tan^2(x)$

   D. $sec^2(x)$

38) The derivative of $f(x) = 2^x$ is:

   A. $ln(2) \cdot 2^x$

   B. e^x

   C. $2x$

   D. $ln(x) \cdot 2^x$

39) Determine the derivative of $f(x) = x \, ln(x)$.

   A. $ln(x) + x$

   B. $x \, ln(x)$

   C. $ln(x)$

   D. $ln(x) + 1$

40) Evaluate: $\int x^2 \, dx$.

   A. $\frac{x^3}{3} + C$

   B. $x^3 + C$

   C. $\frac{x^2}{2} + C$

   D. $x + C$

41) Evaluate: $\int \sqrt{x}\, dx$.

   A. $\frac{2}{3}x^{\frac{3}{2}} + C$

   B. $\frac{1}{2}x^{\frac{3}{2}} + C$

   C. $x^{\frac{1}{2}} + C$

   D. $\frac{3}{2}x^{\frac{3}{2}} + C$

42) Evaluate: $\int (x^3 - 4x)\, dx$.

   A. $\frac{x^4}{4} - 4x^2 + C$

   B. $x^4 - 4x + C$

   C. $\frac{x^4}{4} - 2x^2 + 4 + C$

   D. $\frac{x^4}{4} - 2x^2 + C$

43) Evaluate: $\int sec^2(x)\, dx$.

   A. $sin(x) + C$

   B. $cos(x) + C$

   C. $tan(x) + C$

   D. $sec(x) + C$

44) Evaluate: $\int ln(x)\, dx$.

   A. $ln^2(x) + C$

   B. $x\, ln(x) + C$

   C. $x\, ln(x) + x + C$

   D. $x\, ln(x) - x + C$

45) Evaluate: $\int e^{3x}\, dx$.

   A. $3e^{3x} + C$

   B. $\frac{6}{3}e^{3x} + C$

   C. $e^{3x} + C$

   D. $\frac{1}{3}e^{3x} + C$

46) Evaluate: $\int \frac{x}{x^2+1}\, dx$.

   A. $\frac{1}{2}\ln(x^2+1) + C$

   B. $\arctan(x) + C$

   C. $\frac{1}{2}\ln|x^2+1| + C$

   D. $x\arctan(x) + C$

47) Evaluate: $\int (5x^4 + 3x^2)\, dx$.

   A. $x^5 + x^3 + C$

   B. $x^5 + 3x^2 + C$

   C. $5x^5 + 3x^3 + C$

   D. $5x^5 + x^3 + C$

48) Which of the following is an antiderivative of $f(x) = \frac{1}{x}$?

   A. $\ln|x|$

   B. e^x

   C. $\tan(x)$

   D. $\sin(x)$

49) Evaluate: $\int \frac{dx}{\sqrt{1-x^2}}$.

　A. $arcsin(x) + C$

　B. $arctan(x) + C$

　C. $arccos(x) + C$

　D. $ln|x| + C$

50) Evaluate: $\frac{3-4i}{6+5i}$.

　A. 1

　B. $\frac{1-12i}{61}$

　C. $\frac{-7+39i}{32}$

　D. $\frac{-2-39i}{61}$

51) Evaluate: $\int x^2 e^{x^3} \, dx$.

　A. $e^{x^3} + C$

　B. 1

　C. 0

　D. $\frac{1}{3} e^{x^3} + C$

52) Evaluate: $\int sin^2(x) \, dx$.

　A. $\frac{x}{2} + \frac{sin(2x)}{4} + C$

　B. $\frac{x}{2} - \frac{sin(2x)}{4} + C$

　C. $\frac{x}{2} + \frac{cos(2x)}{4} + C$

　D. $\frac{x}{2} - \frac{cos(2x)}{4} + C$

53) Given the formula $y = -x^2 + 8x - 9$, which one describes its shape?

   A. Ellipse

   B. Hyperbola

   C. Parabola

   D. None of the above

54) Convert the polar coordinate $\left(3, \frac{\pi}{6}\right)$ to Cartesian coordinates.

   A. $\left(\frac{3\sqrt{3}}{2}, \frac{3}{2}\right)$

   B. $\left(\frac{3}{2}, \frac{3\sqrt{3}}{2}\right)$

   C. $\left(-\frac{3\sqrt{3}}{2}, \frac{3}{2}\right)$

   D. $\left(\frac{3}{2}, -\frac{3\sqrt{3}}{2}\right)$

55) The solution to the differential equation $\frac{dy}{dx} = \frac{y}{x}$ is:

   A. $y = ln|x|$

   B. $y = x^2 + C$

   C. $y = Ce^x$

   D. $y = Cx$

56) A tank initially contains 100 liters of pure water. If brine with a salt concentration of 3 grams per liter flows in at a rate of 5 liters per minute, the concentration of salt in the tank after 10 minutes is closest to:

   A. 3 grams per liter

   B. 2 grams per liter

   C. 1.5 grams per liter

   D. 1 gram per liter

57) For the differential equation $y' + y = 0$, the general solution is:

   A. $y = \pm Ce^x$

   B. $y = \pm Ce^{-x}$

   C. $y = xCe^x$

   D. $y = xCe^{-x}$

58) What is the order of the differential equation $y''' + 2y'' + y' = e^x$?

   A. 1

   B. 2

   C. 3

   D. None

59) The differential equation $y' + y^2 = x$ is:

   A. Linear

   B. Non-linear

   C. Homogeneous

   D. Exact

60) Evaluate. $\sum_{n=1}^{7} \left(-\frac{1}{2}\right)^{n-1}$

   A. $\frac{43}{64}$

   B. $\frac{45}{37}$

   C. $\frac{47}{22}$

   D. 1

End of Calculus Practice Test 1

Calculus

Practice Test 2

2024

Total number of questions: 60

Time: <u>No time limit</u>

Calculator is permitted for Calculus Test.

Calculus Practice Test 2 Answer Sheet

Remove (or photocopy) this answer sheet and use it to complete the practice test.

Calculus Practice Test 2 Answer Sheet

1	Ⓐ Ⓑ Ⓒ Ⓓ	21	Ⓐ Ⓑ Ⓒ Ⓓ	41	Ⓐ Ⓑ Ⓒ Ⓓ
2	Ⓐ Ⓑ Ⓒ Ⓓ	22	Ⓐ Ⓑ Ⓒ Ⓓ	42	Ⓐ Ⓑ Ⓒ Ⓓ
3	Ⓐ Ⓑ Ⓒ Ⓓ	23	Ⓐ Ⓑ Ⓒ Ⓓ	43	Ⓐ Ⓑ Ⓒ Ⓓ
4	Ⓐ Ⓑ Ⓒ Ⓓ	24	Ⓐ Ⓑ Ⓒ Ⓓ	44	Ⓐ Ⓑ Ⓒ Ⓓ
5	Ⓐ Ⓑ Ⓒ Ⓓ	25	Ⓐ Ⓑ Ⓒ Ⓓ	45	Ⓐ Ⓑ Ⓒ Ⓓ
6	Ⓐ Ⓑ Ⓒ Ⓓ	26	Ⓐ Ⓑ Ⓒ Ⓓ	46	Ⓐ Ⓑ Ⓒ Ⓓ
7	Ⓐ Ⓑ Ⓒ Ⓓ	27	Ⓐ Ⓑ Ⓒ Ⓓ	47	Ⓐ Ⓑ Ⓒ Ⓓ
8	Ⓐ Ⓑ Ⓒ Ⓓ	28	Ⓐ Ⓑ Ⓒ Ⓓ	48	Ⓐ Ⓑ Ⓒ Ⓓ
9	Ⓐ Ⓑ Ⓒ Ⓓ	29	Ⓐ Ⓑ Ⓒ Ⓓ	49	Ⓐ Ⓑ Ⓒ Ⓓ
10	Ⓐ Ⓑ Ⓒ Ⓓ	30	Ⓐ Ⓑ Ⓒ Ⓓ	50	Ⓐ Ⓑ Ⓒ Ⓓ
11	Ⓐ Ⓑ Ⓒ Ⓓ	31	Ⓐ Ⓑ Ⓒ Ⓓ	51	Ⓐ Ⓑ Ⓒ Ⓓ
12	Ⓐ Ⓑ Ⓒ Ⓓ	32	Ⓐ Ⓑ Ⓒ Ⓓ	52	Ⓐ Ⓑ Ⓒ Ⓓ
13	Ⓐ Ⓑ Ⓒ Ⓓ	33	Ⓐ Ⓑ Ⓒ Ⓓ	53	Ⓐ Ⓑ Ⓒ Ⓓ
14	Ⓐ Ⓑ Ⓒ Ⓓ	34	Ⓐ Ⓑ Ⓒ Ⓓ	54	Ⓐ Ⓑ Ⓒ Ⓓ
15	Ⓐ Ⓑ Ⓒ Ⓓ	35	Ⓐ Ⓑ Ⓒ Ⓓ	55	Ⓐ Ⓑ Ⓒ Ⓓ
16	Ⓐ Ⓑ Ⓒ Ⓓ	36	Ⓐ Ⓑ Ⓒ Ⓓ	56	Ⓐ Ⓑ Ⓒ Ⓓ
17	Ⓐ Ⓑ Ⓒ Ⓓ	37	Ⓐ Ⓑ Ⓒ Ⓓ	57	Ⓐ Ⓑ Ⓒ Ⓓ
18	Ⓐ Ⓑ Ⓒ Ⓓ	38	Ⓐ Ⓑ Ⓒ Ⓓ	58	Ⓐ Ⓑ Ⓒ Ⓓ
19	Ⓐ Ⓑ Ⓒ Ⓓ	39	Ⓐ Ⓑ Ⓒ Ⓓ	59	Ⓐ Ⓑ Ⓒ Ⓓ
20	Ⓐ Ⓑ Ⓒ Ⓓ	40	Ⓐ Ⓑ Ⓒ Ⓓ	60	Ⓐ Ⓑ Ⓒ Ⓓ

EffortlessMath.com

1) Given the 3 functions, find $f\big(g(h(x))\big)$: $\begin{cases} f(x) = \sqrt{2x} \\ g(x) = -4x - 4 \\ h(x) = \frac{x}{3} \end{cases}$.

A. $\dfrac{\sqrt{-8x-8}}{3}$

B. $\sqrt{\dfrac{-8x-24}{3}}$

C. $\sqrt{\dfrac{-8x-8}{3}}$

D. Undefined

2) What is the graph for $f(x) = \lfloor 3x \rfloor$.

A.

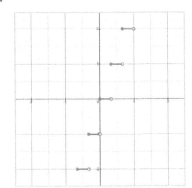

B.

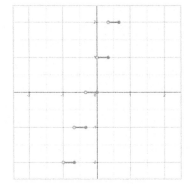

C.

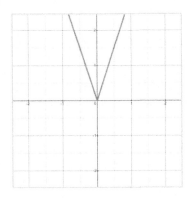

D.

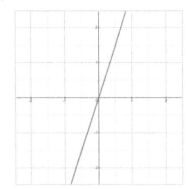

3) Find the inverse of $f(x) = -\pi x^2$.

A. $-\frac{1}{\pi x^2}$

B. πx^2

C. $-\sqrt{-\frac{x}{2\pi}}$

D. $\pm\sqrt{-\frac{x}{\pi}}$

4) Which graph shows $-(x^2 - 5)$?

A.
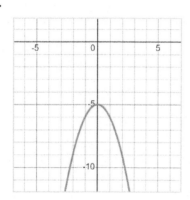

B.
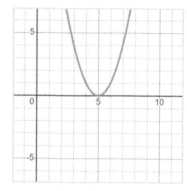

C.
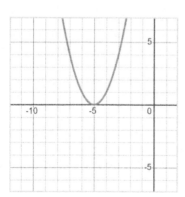

D.
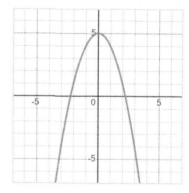

5) If y varies inversely as x and when $x = 7$, $y = 43$, find y when $x = -2$.

   A. 150.5

   B. $\frac{86}{7}$

   C. $-\frac{86}{7}$

   D. -150.5

6) What is the range of $f(x) = \cos\left(\frac{3x}{7}\right) + 1$.

   A. $(-1,1)$

   B. $[-1,1]$

   C. $[0,2]$

   D. $(0,2)$

7) For which θ, $f(x) = (tan(\theta))^x$ is exponential decay?

   A. $0°$

   B. $30°$

   C. $45°$

   D. $60°$

8) Solve $(4\pi)^{\log_{4\pi} 3\pi}$.

   A. 0

   B. π

   C. 3π

   D. 4π

9) Evaluate $log_3 8 + \frac{1}{log_{125} 3}$.

   A. $3 \, log_3 10$

   B. $\frac{1}{3} log_3 10$

   C. 0

   D. 10.5

10) Solve for x: $e^{6x} = 3$.

    A. $\frac{ln \, 6}{3}$

    B. $\frac{ln \, 2}{9}$

    C. $\frac{ln \, x}{3}$

    D. $\frac{ln \, 3}{6}$

11) Given the values for sides of a right triangle, find tan of smallest angle:

    $$4, \, 8, \, \sqrt{80}$$

    A. $\frac{4}{\sqrt{80}}$

    B. $\frac{1}{2}$

    C. $\frac{8}{\sqrt{80}}$

    D. 2

12) Simplify $\frac{sin(2x)}{sec(x)}$.

    A. $2 \, sin(2x)$

    B. $2 \, sin(x) \cdot cos^2(x)$

    C. $2 \, cos(2x)$

    D. $2 \, cos(x)$

13) Given $tan(x) = \frac{1}{2}$, find $tan(2x)$.

   A. $\frac{7}{5}$

   B. $\frac{4}{3}$

   C. 1

   D. $\frac{3}{4}$

14) Simplify $(sin(420°) - sin(300°)) \times 2$.

   A. $2\sqrt{3}$

   B. $\frac{3\sqrt{3}}{2}$

   C. $\sqrt{3}$

   D. $\frac{\sqrt{3}}{2}$

15) What is $\lfloor -2.03 \rfloor - \lfloor 2.2 \rfloor + \lceil 3.5 \rceil - \lceil 4 \rceil$?

   A. -6

   B. -5

   C. -4

   D. -3

16) Given the sequence, find the 8th term: $\left\{0, -\frac{\sqrt{3}}{2}, -\sqrt{3}, \cdots\right\}$.

   A. -12

   B. $-4\sqrt{3}$

   C. $-3.5\sqrt{3}$

   D. 0

17) Find the sum of first 50 natural numbers.

   A. 4,292

   B. 2,550

   C. 5,050

   D. 1,275

18) Find the sum of first 20 terms of this arithmetic series: $\{3, 7, 11, \cdots\}$.

   A. 820

   B. 420

   C. 840

   D. 800

19) Find the sum of first 7 terms: $\{1, \frac{1}{2}, \frac{1}{4}, \ldots\}$.

   A. -1

   B. $\frac{128}{127}$

   C. $\frac{127}{64}$

   D. 2

20) Find the 7th entry on the 10th row of pascal's triangle.

   A. 120

   B. 210

   C. 240

   D. 420

21) What is the midpoint of this interval? $\left(-\frac{3}{7}, \frac{2}{9}\right)$

    A. $-\frac{41}{126}$

    B. $-\frac{13}{126}$

    C. $\frac{13}{126}$

    D. $\frac{41}{126}$

22) Which limit is undefined at $x \to 0$?

    A. $\frac{1}{x}$

    B. $-3x$

    C. $sin(x)$

    D. None

23) If we remove the discontinuity from $\lim_{x \to 1} \frac{x^2-1}{x-1}$, find t: $\begin{cases} \frac{x^2-1}{x-1} & for \ x \neq 1 \\ t & for \ x = 1 \end{cases}$.

    A. 0

    B. 1

    C. 2

    D. 3

24) Using squeeze theorem, find the limit: $\lim_{x \to 0} x \left\lfloor \frac{1}{x} \right\rfloor$.

    A. 1

    B. 2

    C. 3

    D. 4

25) Find the limit: $\lim\limits_{x\to 2^+} \frac{\lfloor x \rfloor - \lfloor x^2 \rfloor}{\lfloor -x \rfloor}$.

   A. $\frac{1}{2}$

   B. $\frac{2}{3}$

   C. $\frac{3}{2}$

   D. 2

26) Evaluate: $\lim\limits_{x\to 0} \frac{\cot(3x)}{\cot(5x)}$.

   A. $\frac{3}{5}$

   B. $\frac{2}{5}$

   C. $\frac{5}{2}$

   D. $\frac{5}{3}$

27) Evaluate: $\lim\limits_{x\to 2} \frac{x^2+2x-8}{x^2-3x+2}$.

   A. Undefined

   B. 4

   C. 5

   D. 6

28) Find the derivative: $(-2x^3)'$.

   A. $-\frac{x^4}{2}$

   B. $6x^{-2}$

   C. $-6x^2$

   D. $\frac{x^{-4}}{2}$

29) Find the derivative: $\left(\dfrac{3x}{-4x^2}\right)'$.

   A. $\dfrac{4x^2}{3}$

   B. $\dfrac{3}{4x^2}$

   C. $-\dfrac{4x^2}{3}$

   D. $-\dfrac{3}{4x^2}$

30) Find the derivative: $(cos'(\pi x))$.

   A. $-sin(\pi x)$

   B. $-\pi \, sin(\pi x)$

   C. $sin(\pi x)$

   D. $\pi \, sin(\pi x)$

31) Given the functions, find the derivative of the composite $f(x) = g\bigl(h(i(x))\bigr)$:

$$\begin{cases} g(h) = 7h - 3 \\ h(i) = 2i^2 + 5 \\ i(x) = -x^2 \end{cases}$$

   A. $56x^3$

   B. $36x^5$

   C. $14x^4 + 32$

   D. ∞

32) Find the derivative: $f(x) = (3x^2 - 2x)^3$.

   A. $270x^5 - 24x^4 + 162x^3 + x^2$

   B. $-144x^5 + 162x^4 - 24x^3 + 270x^2$

   C. $236x^5 + 65x^4 + 83x^3 - 100x^2$

   D. $162x^5 - 270x^4 + 144x^3 - 24x^2$

33) Find the derivative: $\left(\sqrt{(-2x+5)}\right)'$.

   A. $2\sqrt{(-2x+5)}$

   B. $\dfrac{1}{2\sqrt{(-2x+5)}}$

   C. $-\dfrac{1}{\sqrt{(-2x+5)}}$

   D. $-\dfrac{2}{\sqrt{(-2x+5)}}$

34) Find the derivative of $f(x) = \sqrt{\cos(2x) - 4}$.

   A. $\dfrac{1}{2\sqrt{\cos(2x)-4}}$

   B. $-\dfrac{\sin(2x)}{\sqrt{\cos(2x)-4}}$

   C. $\dfrac{\sin(2x)}{\sqrt{\cos(2x)-4}}$

   D. $-\dfrac{\sin(2x)}{2\sqrt{\cos(2x)-4}}$

35) Find the derivative of $\log_3 3x$.

   A. $\dfrac{1}{3x \ln 3}$

   B. $\dfrac{1}{x \ln 3}$

   C. $\dfrac{1}{x \ln 3}$

   D. $\dfrac{1}{3 \ln 3}$

36) Find the second derivative of $f(x) = 9x^3 - 4x$.

   A. $54x$

   B. $27x^2 - 4$

   C. 0

   D. 54

37) Find $f'''(x)$: $f(x) = cos(x) + 3x^2$.

　　A. $sin(x)$

　　B. $cos(x)$

　　C. $tan(x)$

　　D. $cot(x)$

38) If $x^2 + y^2 = 25$, find $\frac{dy}{dx}$.

　　A. $\frac{x}{y}$

　　B. $-\frac{x}{y}$

　　C. $2x + 2y$

　　D. $-\frac{2x}{y}$

39) If $x^2 y = 6$, find $\frac{dy}{dx}$.

　　A. $-\frac{12}{x^3}$

　　B. $\frac{12}{x^3}$

　　C. $-\frac{6}{x^2}$

　　D. $\frac{6}{x^2}$

40) Which one Can be the integral of $36x + 6$?

　　A. $18x^2 + 6x - 1$

　　B. $18x^2 + 6x - 3$

　　C. $18x^2 - 6x - 1$

　　D. Choice A and B

41) Find: $\int_1^4 6x^3\, dx$.

   A. 378

   B. 378.5

   C. 382

   D. 382.5

42) In solving optimization problems with derivatives, what is the purpose of finding critical points?

   A. To determine the function's continuity

   B. To find where the function is undefined.

   C. To locate potential maximum or minimum values of the function

   D. To integrate the function

43) Calculate the midpoint Riemann sum approximation for the integral of $f(x) = x^2$ over the interval [1,3] using two subintervals.

   A. 8.5

   B. 10

   C. 12

   D. 14

44) Find the sum of first 3 terms of this geometric series: $a_4 = \frac{-27}{2}$, $a_5 = 4.5$.

   A. 384

   B. 283.5

   C. 363.5

   D. 383.5

45) Evaluate the geometric series described as: $\sum_{i=1}^{\infty} \left(-\frac{2}{3}\right)^{i-1}$.

   A. $\frac{5}{3}$

   B. $\frac{4}{3}$

   C. $\frac{3}{5}$

   D. $\frac{2}{5}$

46) Evaluate $\int \sin^2(x)\, dx$.

   A. $x - \frac{1}{2}\sin(2x) + C$

   B. $\frac{1}{2}x - \frac{1}{4}\cos(2x) + C$

   C. $\frac{1}{2}x - \frac{1}{4}\sin(2x) + C$

   D. $x - \frac{1}{2}\cos(2x) + C$

47) Evaluate: $\int x^2 \cos(x)\, dx$.

   A. $x^2 \cos(x) + 2x \sin(x) - 2\cos(x) + C$

   B. $\cos(x) + x + C$

   C. $x \sin(x) + x \cos(x) + C$

   D. $x^2 \sin(x) + 2x \cos(x) - 2\sin(x) + C$

48) Find $\int \cos(3x)\, dx$.

   A. $\frac{1}{3}\sin(3x) + C$

   B. $-\frac{1}{3}\sin(3x) + C$

   C. $3\cos(3x) + C$

   D. $\frac{1}{3}\cos(3x) + C$

49) Find $\int \sqrt{5x}\, dx$.

   A. $\frac{2}{3}\sqrt{x^3}$

   B. $\frac{2}{3}\sqrt{x^3} + C$

   C. $\frac{2\sqrt{5}}{3}\sqrt{x^3} + C$

   D. $\frac{2\sqrt{5}}{5}\sqrt{x^3} + C$

50) Solve this improper integral $\int_1^\infty \frac{1}{x^3}\, dx$.

   A. -1

   B. $-\frac{1}{2}$

   C. $\frac{1}{2}$

   D. 1

51) Which differential equation is 2nd order and of 3rd degree?

   A. $-\frac{d^2x}{dx^2} + \left(\frac{dy}{dx}\right)^3 - y^2 = 0$

   B. $\left(\frac{d^2y}{dx^2}\right)^2 + \left(\frac{dy}{dx}\right)^3 = 0$

   C. $\left(\frac{d^3y}{dx^3}\right)^2 - \frac{dy}{dx} + 2y^3 = 0$

   D. $\left(\frac{d^2y}{dx^2}\right)^3 = 0$

52) Find the differential equation that is homogenous, linear, of 2nd degree and is 1st order.

A. $sin(y) + \frac{d^2y}{dx^2} = 0$

B. $y\frac{dy}{dx} + xy = 0$

C. $-2xy + \left(\frac{dy}{dx}\right)^2 = 0$

D. $sin(x) + \left(\frac{d^3y}{dx^3}\right)^2 = 0$

53) Given the slope field, what could be the answer to $\frac{dy}{dx}$?

A. $3x$

B. $3x - 5$

C. $-x$

D. $-3x$

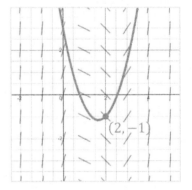

54) Nobelium element has a half-life of 58 minutes. If you start with 100 grams of nobelium, which differential equation would describe the amount $N(t)$ of nobelium remaining after t minutes?

A. $\frac{dN}{dt} = \frac{-\ln 2}{58} \times N(t)$

B. $\frac{dN}{dt} = \frac{\ln 2}{58} \times N(t)$

C. $\frac{dN}{dt} = \frac{-58}{\ln 2} \times N(t)$

D. $\frac{dN}{dt} = \frac{58}{\ln 2} \times N(t)$

55) Convert the polar coordinate $\left(5, \frac{\pi}{3}\right)$ to rectangular coordinates.

   A. $\left(\frac{5}{2}, \frac{5\sqrt{3}}{2}\right)$

   B. $\left(\frac{5\sqrt{3}}{2}, \frac{5}{2}\right)$

   C. $\left(\frac{5\sqrt{2}}{2}, \frac{5\sqrt{2}}{2}\right)$

   D. $\left(5\sqrt{3}, 5\right)$

56) Evaluate: $(3 + 9i) \cdot (-2 - 6i)$.

   A. $48 - 36i$

   B. $36 + 24i$

   C. $24 - 12i$

   D. $12 + 6i$

57) Evaluate: $\frac{4+3i}{7-i}$.

   A. $\frac{i}{2}$

   B. $\frac{i+1}{2}$

   C. $\frac{i+1}{2}$

   D. 0

58) Find the derivative: $\left(\frac{\sin(3x)}{\sqrt{x^3}}\right)'$.

   A. $\frac{\cos(3x) \cdot \sqrt{x^3} - \sqrt{x} \sin(3x)}{x^3}$

   B. $\frac{3\cos(3x) \cdot \sqrt{x^3} - \frac{3}{2}\sqrt{x} \sin(3x)}{x^3}$

   C. $\frac{\cos(3x) \cdot \sqrt{x^3} - \frac{3}{2} \sin(3x)}{x^3}$

   D. $3\cos(3x) \cdot \sqrt{x^3}$

59) What formula is this?

$$\binom{n}{0}x^n y^0 + \binom{n}{1}x^{n-1}y^1 + \cdots + \binom{n}{n-1}x^1 y^{n-1} + \binom{n}{n}x^0 y^n$$

A. Ellipse

B. Euler's method

C. Binomial theorem

D. None

60) Which one shows the main application of integrals?

A. Instant change

B. Total accumulation

C. Tangent line

D. All options

End of Calculus Practice Test 2

Calculus Practice Tests Answer Keys

Now, it's time to review your results to see where you went wrong and what areas you need to improve.

Calculus Practice Test 1							Calculus Practice Test 2						
1	C	21	B	41	A		1	B	21	B	41	D
2	B	22	D	42	D		2	A	22	A	42	C
3	B	23	C	43	C		3	D	23	C	43	A
4	A	24	A	44	D		4	D	24	A	44	B
5	B	25	C	45	D		5	D	25	C	45	C
6	A	26	D	46	A		6	C	26	D	46	C
7	B	27	C	47	A		7	B	27	D	47	D
8	D	28	B	48	A		8	C	28	C	48	A
9	B	29	B	49	A		9	A	29	B	49	C
10	B	30	A	50	D		10	D	30	B	50	C
11	D	31	B	51	D		11	B	31	A	51	D
12	A	32	A	52	B		12	B	32	D	52	C
13	B	33	A	53	C		13	B	33	C	53	B
14	B	34	C	54	A		14	A	34	B	54	A
15	C	35	A	55	D		15	B	35	B	55	A
16	B	36	B	56	B		16	C	36	A	56	A
17	B	37	A	57	B		17	D	37	A	57	C
18	B	38	A	58	C		18	A	38	B	58	B
19	A	39	D	59	B		19	C	39	A	59	C
20	B	40	A	60	A		20	B	40	D	60	B

Calculus Practice Tests Answers and Explanations

Calculus Practice Tests 1 Explanations

1) **Choice C is correct.**

   The expression inside the square root, $x - 5$, must be non-negative for the function to have real values. Therefore, $x - 5 \geq 0$ or $x \geq 5$.

2) **Choice B is correct.**

   The absolute value function has a sharp corner at $x = 0$, so its derivative does not exist there.

3) **Choice B is correct.**

   First, find $g(3) = 3^2 = 9$. Now, $f(g(3)) = f(9) = 2(9) + 1 = 19$.

4) **Choice A is correct.**

   This is a property of inverse functions. If you plug the inverse function into the original function, you get back your input. So, the input or a is the final answer.

5) **Choice B is correct.**

   To find the inverse of $y = x^2$ for $x \geq 0$, we need to switch places of x and y.

   Switching the x and y, we have: $x = y^2$, then taking the square root of both sides: $y = \sqrt{x}$.

6) **Choice A is correct.**

   $f(x - 3)$ shifts the graph of x^2 three units to the right, and $+1$ shifts it one unit up.

7) **Choice B is correct.**

   The negative sign flips the graph vertically, and the vertical axis is y-axis.

Calculus for Beginners

8) Choice D is correct.

A sequence is arithmetic if the difference between consecutive terms is constant.

3, 6, 12, 24, $\cdots$: The differences are 3, 6, 12, $\cdots$ which are not constant.

2, 5, 9, 13, $\cdots$: The differences are 3, 4, 4, $\cdots$ which are not constant.

1, 4, 9, 16, $\cdots$: The differences are 3, 5, 7, $\cdots$ which are not constant.

5, 3, 1, -1, $\cdots$: The differences are $-2, -2, -2, \cdots$ which are constant.

9) Choice B is correct.

$2n + 3$ is the general formula for the nth term. Plug $n = 5$ into the expression to find the 5th term: $a_5 = 2(5) + 3 = 13$.

10) Choice B is correct.

A sequence is geometric if the ratios between consecutive terms is constant.

1, 3, 5, 7, $\cdots$: The ratios are $3, \frac{5}{3}, \frac{7}{5}, \cdots$ which are not constant.

4, 2, 1, 0.5, $\cdots$: The ratios are $\frac{1}{2}, \frac{1}{2}, \frac{1}{2}, \cdots$ which are constant.

1, 1, 2, 2, $\cdots$: The ratios are $1, 2, 1, \cdots$ which are not constant.

2, 6, 12, 20, $\cdots$: The ratios are $3, 2, \frac{5}{3}, \cdots$ which are not constant.

11) Choice D is correct.

Using formula for the n-th term of an arithmetic sequence: $a_n = a_1 + (n-1)d$. Substitute $n = 6$, then: $a_6 = 3 + (6-1)4 = 3 + 5 \times 4 = 23$. So,

$$\sum_{n=1}^{n} a_n = \frac{n}{2}(a_1 + a_n) \Rightarrow \sum_{n=1}^{6} a_n = \frac{6}{2}(3 + 23) = 3(26) = 78$$

Calculus Practice Tests 1 Explanations

12) Choice A is correct.

We could use the formula to find the sum, but since we only need the sum of first 3 terms, it's easier to find the terms using the definition of geometric sequences, then sum the terms: $S_3 = 3 + 3(2) + 3(2^2) = 3 + 6 + 12 = 21$.

13) Choice B is correct.

Given the first term $a_1 = 4$, and the sum of all terms 8:

The sum of an infinite geometric series is $S = \frac{a_1}{1-r}$.

$8 = \frac{4}{1-r} \Rightarrow 1 - r = \frac{1}{2} \Rightarrow -r = \frac{1}{2} - 1 = -\frac{1}{2} \Rightarrow r = \frac{1}{2} = 0.5$.

14) Choice B is correct.

The series $1 - \frac{1}{2} + \frac{1}{4} - \frac{1}{8} + \cdots$, being an alternating geometric series with $|r| < 1$, is convergent, since the terms of this series get smaller and smaller as n gets bigger.

15) Choice C is correct.

For x^2 being a continuous function, we can find the limit at 2 by direct substitution: $\lim_{x \to 2} x^2 = f(2) = 2^2 = 4$.

16) Choice B is correct.

For $2x - 5$ being a continuous function, we can find the limit at 3 by directly substituting 3 for x: $\lim_{x \to 3}(2x - 5) = 2(3) - 5 = 6 - 5 = 1$.

17) Choice B is correct.

The function simplifies to $f(x) = \frac{(x-3)(x+3)}{x-3} = x + 3$, when you factor out the numerator and cancel $(x - 3)$. However, it's undefined at $x = 3$, making it also discontinuous at this point.

18) **Choice B is correct.**

First, we put $x = 0$ in the equations of the numerator and denominator functions. So, $\lim\limits_{x \to 0} \dfrac{sin(x)}{x} = \dfrac{sin(0)}{0} = \dfrac{0}{0}$. We have an indeterminate form of $\dfrac{0}{0}$.

Now, according to these numerator and denominator functions, i.e. $y = sin(x)$ and $y = x$ are continuous and derivable, we can use Hopital's Rule and calculate the limit. Applying Hopital's Rule, we take the derivatives of the numerator and denominator separately. So: $y = sin(x) \Rightarrow y' = cos(x)$ and $y = x \Rightarrow y' = 1$. Next, we have:

$$\lim_{x \to 0} \frac{sin(x)}{x} = \lim_{x \to 0} cos(x) = cos(0) = 1 \Rightarrow \lim_{x \to 0} \frac{sin(x)}{x} = 1$$

19) **Choice A is correct.**

$f(x)$ is continuous for all real values, so we can use direct substitution:

$\lim\limits_{x \to 1}(x^3 - x) = 1^3 - 1 = 1 - 1 = 0.$

20) **Choice B is correct.**

The square root function is defined and continuous for all non-negative real numbers. So, x must be equal or greater than zero, so: $x \geq 0$.

21) **Choice B is correct.**

From the unit circle, we know that every sin value is between 1 and -1. So: $-1 \leq sin\left(\dfrac{1}{x}\right) \leq 1$, then multiplying all by x^2: $-x^2 \leq x^2 sin\left(\dfrac{1}{x}\right) \leq x^2$, and using the squeeze theorem:

As $x \to 0$, both x^2 and $-x^2$ approach 0, so $x^2 sin\left(\dfrac{1}{x}\right)$ must approach 0 too.

22) **Choice D is correct.**

If x is slightly smaller than 2, then $x - 2$ is slightly smaller than 0, so:

$\dfrac{1}{0^-} = -\infty.$

23) Choice C is correct.

As we approach 0 from right on the graph of $\frac{1}{x}$, the result approaches infinity.

24) Choice A is correct.

At $x = 0$, we encounter $\frac{0}{0}$ and that's the point of discontinuity for $\frac{x}{|x|}$. Because, for $x \geq 0$, then $|x| = x$. So, $\frac{x}{|x|} = \frac{x}{x} = 1$ and for $x \leq 0$, then $|x| = -x$. So, $\frac{x}{|x|} = \frac{x}{-x} = -1$. Therefore, $\lim_{x \to 0^-} \frac{x}{|x|} = \lim_{x \to 0^-} -1 = -1$ and $\lim_{x \to 0^+} \frac{x}{|x|} = \lim_{x \to 0^+} 1 = 1$.

25) Choice C is correct.

Continuity at a point requires the function to be defined there, the limit to exist, and the function's value to equal its limit.

26) Choice D is correct.

$f(x) = x^2 + cx + 4$ is a quadratic function and quadratic functions are continuous everywhere, regardless of the value of c.

27) Choice C is correct.

This is a factorable expression: $\frac{x^2-1}{x-1} = \frac{(x+1)(x-1)}{x-1} = x + 1$, so the limit as $x \to 1$ is: $\lim_{x \to 1} \frac{x^2-1}{x-1} = \lim_{x \to 1} (x+1) = 1 + 1 = 2$.

28) Choice B is correct.

When $x = 2$, the denominator becomes 0. Thus, $x = 2$ is a point of discontinuity for the function. However, it's a removable discontinuity since factoring the numerator will cancel out the $x - 2$ term.

29) **Choice B is correct.**

In order to solve this problem, when $x \to 0^+$, it can be assumed by setting $x = \frac{1}{y}$ or $y = \frac{1}{x}$. So, $x \ln(x) = \frac{1}{y} \ln\left(\frac{1}{y}\right) = \frac{\ln\left(\frac{1}{y}\right)}{y}$, that $x \to 0^+ \Rightarrow y = \frac{1}{x} \to \frac{1}{0^+} = +\infty$. Next when $y \to +\infty$, we have:

$$\lim_{x \to 0^+} x \ln(x) = \lim_{y \to +\infty} \frac{\ln\left(\frac{1}{y}\right)}{y}$$

Since the functions y and $\ln\left(\frac{1}{y}\right)$ are continuous on the interval $y > 0$, so by deriving these functions and calculating the limit, we have:

$$\lim_{y \to +\infty} \frac{\ln\left(\frac{1}{y}\right)}{y} = \lim_{y \to +\infty} \frac{\left(\ln\left(\frac{1}{y}\right)\right)'}{(y)'} = \lim_{y \to +\infty} \frac{-\frac{1}{y^2} \times y}{1} = \lim_{y \to +\infty} -\frac{1}{y} = -\frac{1}{+\infty} = 0$$

30) **Choice A is correct.**

As x approaches $\frac{\pi}{2}$ from left, the $\sin x$ component of $\tan x$ approaches 1, such that $\tan(x) = \frac{\sin(x)}{\cos(x)}$ and $x \to \frac{\pi}{2} \Rightarrow \sin(x) \to 1$, while the $\cos(x)$ approaches 0^+, making the tangent function increase without bound. Thus, the limit is positive infinity.

$$\lim_{x \to \frac{\pi}{2}^-} \tan(x) = \lim_{x \to \frac{\pi}{2}^-} \frac{\sin(x)}{\cos(x)} = \frac{1}{0^+} = +\infty$$

31) **Choice B is correct.**

Using the product rule: $(x^2)' \cdot e^x + x^2 \cdot (e^x)' = 2xe^x + x^2 e^x$.

32) **Choice A is correct.**

Chain rule is used for the composition of functions.

33) **Choice A is correct.**

Using the power rule, we have: $f'(x) = (x^2)' - (4x)' + (7)' = 2x - 4$.

Calculus Practice Tests 1 Explanations

34) Choice C is correct.

Using chain rule, we find the derivative of inner function:

$(5x)' = 5$, then, using the formula $\left(ln(f(x))\right)' = \frac{1}{f(x)} f'(x)$:

$$y'(x) = \frac{1}{5x} \cdot 5 = \frac{1}{x}$$

35) Choice A is correct.

This is a well-known identity to find the derivative of inverse of sine:

$$f'(arcsin(x)) = \frac{1}{\sqrt{1-x^2}}$$

36) Choice B is correct.

Using the chain rule, $\frac{dy}{dx} = f'(g(x)) \times g'(x)$. Since $f(x) = 5e^x$ and $g(x) = 2x$, we get: $f(g(x)) = 5e^{2x}$. Now, we have: $(5e^{2x})' = (5e^{2x})' \times (2x)'$. Next, we know that $\left(c \cdot h(x)\right)' = c \cdot h'(x)$. So, $(5e^x)' = 5 \cdot (e^x)$. Therefore,

$$(5e^{2x})' = (5e^{2x})' \times (2x)' = 5e^{2x} \times 2 = 10e^{2x}$$

37) Choice A is correct.

We know that $sec(x) = \frac{1}{cos(x)}$, we'll use the quotient rule:

$$y' = \frac{0 \cdot cos(x) - 1 \cdot (-sin(x))}{cos^2(x)} = \frac{sin(x)}{cos^2(x)} = \frac{1}{cos(x)} \times \frac{sin(x)}{cos(x)} \Rightarrow y' = sec(x)\, tan(x)$$

38) Choice A is correct.

Using the formula $(a^x)' = a^x \cdot ln(a)$: $(2^x)' = ln(2) \cdot 2^x$.

39) Choice D is correct.

Using the product rule:

$$(x)' \cdot (ln(x)) + (x) \cdot (ln(x)) = 1 \cdot ln(x) + x \cdot \frac{1}{x} = ln(x) + 1$$

40) **Choice A is correct.**

Using the power rule: $\int x^2 \, dx = \frac{x^3}{3} + C$.

41) **Choice A is correct.**

Using the power rule:

$$\int \sqrt{x} \, dx = \int x^{\frac{1}{2}} \, dx = \frac{x^{\frac{1}{2}+1}}{\frac{1}{2}+1} = \frac{x^{\frac{3}{2}}}{\frac{3}{2}} = \frac{2}{3} x^{\frac{3}{2}} + C$$

42) **Choice D is correct.**

Using the power rule:

$$\int (x^3 - 4x) \, dx = \int x^3 \, dx - \int 4x \, dx = \left(\frac{x^4}{4} + C_1\right) - \left(4 \cdot \frac{x^2}{2} + C_2\right)$$

$$= \frac{x^4}{4} - 2x^2 + (C_1 - C_2) = \frac{x^4}{4} - 2x^2 + C$$

43) **Choice C is correct.**

Using substitution, we assume $u = tan(x)$, and using quotient rule of derivatives: $\frac{du}{dx} = \frac{1}{cos^2(x)} \Rightarrow du = \frac{1}{cos^2(x)} dx \Rightarrow du = sec^2(x) \, dx$,

$$\int sec^2(x) \, dx = \int du = u + C = tan(x) + C$$

44) **Choice D is correct.**

To solve, we consider the integral as two parts.

$$\int ln(x) \, dx = \int ln(x) \cdot 1 \, dx = \int (ln(x)) \cdot (1 \, dx)$$

Let $u = ln(x) \Rightarrow du = \frac{1}{x} dx$, and $dv = 1 \, dx \Rightarrow v = x$. So, we have:

$$\int \ln(x)\, dx = \int u \cdot dv = u \cdot v - \int v \cdot du = \ln(x) \cdot x - \int x \cdot \frac{1}{x} dx$$

$$= x\ln(x) - \int 1\, dx = x\ln(x) - x + C$$

45) Choice D is correct.

Using the formula: $\int e^{ax}\, dx = \frac{1}{a} e^{ax} + C$: $\int e^{3x}\, dx = \frac{1}{3} e^{3x} + C$.

46) Choice A is correct.

Using substitution rule, let u be $x^2 + 1$, which has a derivative of:

$du = 2x\, dx$, meaning $\frac{du}{2} = x\, dx$ and by substituting this into the main integral, we have:

$$\int \frac{x}{x^2+1} dx = \int \frac{1}{u} \cdot \frac{1}{2} du = \frac{1}{2} \int \frac{1}{u} du = \frac{1}{2} \ln|u| + C = \frac{1}{2} \ln|x^2 + 1| + C$$

And since $x^2 + 1$ is always positive, we can have: $\frac{1}{2} \ln(x^2 + 1) + C$.

47) Choice A is correct.

$$\int (5x^4 + 3x^2)\, dx = \frac{5x^5}{5} + \frac{3x^3}{3} + C = x^5 + x^3 + C$$

48) Choice A is correct.

The formula to find the antiderivative of $\frac{1}{x}$, is: $\int \frac{1}{x} dx = \ln|x| + C$.

49) Choice A is correct.

The formula to find the derivative of inverse sine function is:

$$(\arcsin(x))' = \frac{1}{\sqrt{1-x^2}}$$

So, the integral of $\frac{1}{\sqrt{1-x^2}}$ is $\arcsin(x) + C$.

Calculus for Beginners

50) **Choice D is correct.**

To evaluate, $\frac{3-4i}{6+5i}$ needs to be multiplied by a proper 1 (conjugate is $6-5i$), being $\frac{6-5i}{6-5i}$: $\frac{3-4i}{6+5i} \times \frac{6-5i}{6-5i} = \frac{18-15i-24i+20i^2}{6^2-(5i)^2} = \frac{18-39i+20(-1)}{36-25(-1)} = \frac{18-39i-20}{36+25} = \frac{-2-39i}{61}$.

51) **Choice D is correct.**

To integrate $\int x^2 e^{x^3} dx$, we can use the substitution $u = x^3$, which gives us $\frac{du}{dx} = 3x^2$ then, $x^2 dx = \frac{du}{3}$. Substituting this into the integral, we get:

$$\int x^2 e^{x^3} dx = \int e^{x^3} x^2 dx = \int e^u \cdot \frac{1}{3} du = \int \frac{1}{3} e^u du = \frac{1}{3} e^u + C = \frac{1}{3} e^{x^3} + C$$

52) **Choice B is correct.**

To find the integral of $\int \sin^2(x) dx$, we first use formula $\sin^2(x) = \left(\frac{1-\cos(2x)}{2}\right)$. Therefore, by substituting we have:

$$\int \sin^2(x) dx = \int \frac{1-\cos(2x)}{2} dx = \frac{1}{2} \int (1 - \cos(2x)) dx$$

$$= \frac{1}{2}\left[\int 1 dx - \int \cos(2x) dx\right] = \frac{1}{2} \int 1 dx - \frac{1}{2} \int \cos(2x) dx$$

First part gives: $\frac{1}{2} \int 1 dx = \frac{1}{2} x + C_1$.

For the second part, we use $\int \cos(nx) dx = \frac{1}{n} \sin(nx) + C$.

$$\frac{1}{2} \int \cos(2x) dx = \frac{1}{2}\left(\frac{1}{2} \sin(2x)\right) + C = \frac{1}{4} \sin(2x) + C_2$$

When combined:

$$\int \sin^2(x) dx = \frac{1}{2} x - \frac{1}{4} \sin(2x) + C = \frac{x}{2} - \frac{\sin(2x)}{4} + C$$

53) **Choice C is correct.**

The formula is in the $y = ax^2 + bx + c$ form, so it is a parabola.

54) **Choice A is correct.**

Using the formulas $x = r\cos(\theta)$ and $y = r\sin(\theta)$, we have $r = 3$ and $\theta = \frac{\pi}{6}$. So: $x = 3\cos\left(\frac{\pi}{6}\right) = \frac{3\sqrt{3}}{2}$ and $y = 3\sin\left(\frac{\pi}{6}\right) = \frac{3}{2}$. Therefore, $\left(\frac{3\sqrt{3}}{2}, \frac{3}{2}\right)$.

55) **Choice D is correct.**

By separating variables and integrating, we get:

$$\frac{dy}{dx} = \frac{y}{x} \Rightarrow \frac{dy}{y} = \frac{dx}{x} \Rightarrow \int \frac{dy}{y} = \int \frac{dx}{x} \Rightarrow ln|y| + K_1 = ln|x| + K_2$$

$$\Rightarrow ln|y| = ln|x| + (K_2 - K_1) \Rightarrow ln|y| = ln|x| + K$$

We assume C so that K is equal to $ln\,C$. We have the following:

$$ln|y| = ln|x| + K \Rightarrow ln|y| = ln|x| + ln\,C$$

Next, by using the formula of the sum of logarithms $log_a bc = log_a b + log_a c$, we get: $ln|y| = ln|x| + ln\,C = ln\,C|x| \Rightarrow y = Cx$.

Where K_1, K_2, K, and $ln\,C$ are real numbers.

56) **Choice B is correct.**

After 10 minutes, the tank would have $5 \times 10 = 50$ liters of liquids (and if we ignore the volume of salt in the 50 liters of brine).

On the other hand, 3 grams per liter of salt in the initial solution for 100 liters of water are $3 \times 100 = 300$ grams of salt. Therefore, the concentration of 300 grams of salt in $100 + 50 = 150$ liters of water become: $\frac{300}{150} = 2$ grams per liter.

57) **Choice B is correct.**

$y' + y = 0 \Rightarrow y' = -y$, so $\frac{dy}{dx} = -y$ or $\frac{dy}{y} = -dx$. Then, integrating both sides:

$$\int \frac{dy}{y} = \int -dx \Rightarrow ln|y| = -x + C_1$$

Now, to solve for y: exponentiate both sides to get rid of the logarithm.

$|y| = e^{-x+C} \Rightarrow |y| = e^{-x} \cdot e^C$, and since e^C is just a constant, we can call it C:

$$|y| = Ce^{-x} \Rightarrow y = \pm Ce^{-x}$$

So, the general solution will be: $y = \pm Ce^{-x}$.

58) Choice C is correct.

The order of the equation is 3. Since y''' is the highest derivative in the equation.

59) Choice B is correct.

According to the differential equation $y' + y^2 = x$. The presence of y^2 makes it non-linear.

60) Choice A is correct.

Using the formula $\sum_{i=1}^{n} a_1 r^{i-1}$, we find that the first term in $\sum_{n=1}^{7} \left(-\frac{1}{2}\right)^{n-1}$ is 1 and the ratio is $-\frac{1}{2}$, then using the formula $a_1 \left(\frac{1-r^n}{1-r}\right)$, we can calculate the sum of the first 7 terms of this geometric sequence:

$$S_7 = \sum_{n=1}^{7} \left(-\frac{1}{2}\right)^{n-1} = (1)\left(\frac{1-\left(-\frac{1}{2}\right)^7}{1-\left(-\frac{1}{2}\right)}\right) \Rightarrow S_7 = \left(\frac{1+\frac{1}{128}}{1+\frac{1}{2}}\right) = \left(\frac{\frac{129}{128}}{\frac{3}{2}}\right) = \frac{43}{64}$$

Calculus Practice Tests 2 Explanations

1) **Choice B is correct.**

   Starting from the inner function and working our way out:

   $$g(h(x)) = -4\left(\frac{x}{3}\right) - 4 = \frac{-4x-12}{3}$$

   $$f(g(h(x))) = f\left(\frac{-4x-12}{3}\right) = \sqrt{2\left(\frac{-4x-12}{3}\right)} = \sqrt{\frac{-8x-24}{3}}$$

2) **Choice A is correct.**

   Let's choose some random points for the x and find $f(x)$ for them:

x	0	0.3	0.5	-0.5
$f(x)$	0	0	1	-2

   Which looks like the graph of option A.

   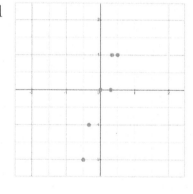

   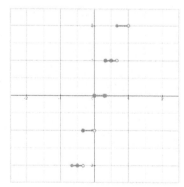

3) **Choice D is correct.**

   To find the inverse of this function $y = -\pi x^2$, we switch the placement of x and y and solve for the new y: $x = -\pi y^2 \Rightarrow -\frac{x}{\pi} = y^2 \Rightarrow y = \pm\sqrt{-\frac{x}{\pi}}$.

4) **Choice D is correct**

   We can rewrite the expression as $-x^2 + 5$, so we have a horizontally flipped x^2 Graph, which will be moved up by 5 units.

   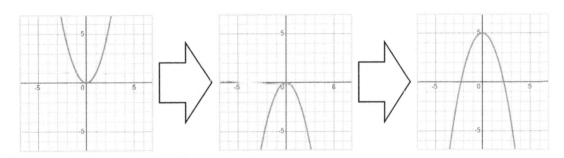

5) **Choice D is correct.**

   Since y varies inversely as x is: $xy = k$. So, we have:

   $x_1 = 7$, and $y_1 = 43 \Rightarrow x_1 \cdot y_1 = 7 \times 43 = 301 \Rightarrow k = 301$.

   Therefore,

   $x_2 = -2$, and $k = 301 \Rightarrow x_2 \cdot y_2 = k \Rightarrow -2y_2 = 301 \Rightarrow y_2 = -\frac{301}{2} = -150.5$.

6) **Choice C is correct.**

   For the function $cos\left(\frac{3x}{7}\right)$ can yield results between -1 and 1, including those points themselves: $-1 \leq cos\left(\frac{3x}{7}\right) \leq 1$. So, the range of $f(x)$ is:

   $$-1 \leq cos\left(\frac{3x}{7}\right) \leq 1 \Rightarrow -1 + 1 \leq cos\left(\frac{3x}{7}\right) + 1 \leq 1 + 1 \Rightarrow 0 \leq f(x) \leq 2$$

   It means that, $[0, 2]$.

7) **Choice B is correct.**

   For a^x to be exponential decay, a must be $0 < a < 1$, and among the options, the only angle with a tangent smaller than 1 and bigger than 0, is: $tan\ 30° \approx 0.6$. So, for $\theta = 30°$, $f(x)$ is exponential decay.

8) **Choice C is correct.**

Using $a^{\log_a k} = k$ identity, the answer is 3π, because 4π is raised to the power of a logarithm with the base of also 4π, so the expression of logarithm will be the answer.

9) **Choice A is correct.**

Using the power rule and the identity $\log_a x = \frac{1}{\log_x a}$, we have:

$$\log_3 8 + \frac{1}{\log_{125} 3} = \log_3 2^3 + \log_3 5^3 = 3\log_3 2 + 3\log_3 5 = 3\log_3 10$$

10) **Choice D is correct.**

Taking the ln of both sides, we have:

$$\ln e^{6x} = \ln 3 \Rightarrow 6x \ln e = \ln 3 \Rightarrow x = \frac{\ln 3}{6}$$

11) **Choice B is correct.**

Our triangle looks like this:

And the smallest angle is shown in red.

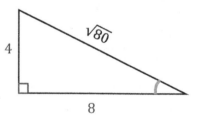

$$\tan(\theta) = \frac{opposite}{adjacent} = \frac{4}{8} = \frac{1}{2}$$

12) **Choice B is correct.**

By $sin(2x)$ formula: $sin(2x) = 2\sin(x)\cos(x)$, we have:

$$\frac{\sin(2x)}{\sec(x)} = \frac{2\sin(x)\cos(x)}{\frac{1}{\cos(x)}} = 2\sin(x) \cdot \cos^2(x)$$

13) **Choice B is correct.**

We have $tan(x)$ and need to find $tan(2x)$ with it, so using the trigonometric identity $tan(2x) = \frac{2\,tan(x)}{1-tan^2(x)} \Rightarrow tan(2x) = \frac{2 \times \frac{1}{2}}{1-\left(\frac{1}{2}\right)^2} = \frac{1}{1-\frac{1}{4}} = \frac{1}{\frac{3}{4}} = \frac{4}{3}.$

14) **Choice A is correct.**

$sin(420°) = sin(2\pi + 60°) = sin(60°)$

$sin(300°) = sin(2\pi + (-60°)) = sin(-60°) = -sin(60°)$

Therefore,

$(sin(420°) - sin(300°)) \times 2 = (sin(60°) - (-sin(60°))) \times 2$

$= (2\,sin(60°)) \times 2 = 4\,sin(60°) = \frac{4\sqrt{3}}{2} = 2\sqrt{3}$

15) **Choice B is correct.**

$\left.\begin{array}{l} -3 \leq -2.03 < -2 \Rightarrow \lfloor -2.03 \rfloor = -3 \\ 2 \leq 2.2 < 3 \Rightarrow \lfloor 2.2 \rfloor = 2 \\ 3 < 3.5 \leq 4 \Rightarrow \lceil 3.5 \rceil = 4 \\ \lceil 4 \rceil = 4 \end{array}\right\} \Rightarrow -3 - 2 + 4 - 4 = -5$

16) **Choice C is correct.**

This is an arithmetic sequence, with $a = 0$ and $d = -\frac{\sqrt{3}}{2}$, so:

$a_8 = a + (8-1)d = 0 + 7\left(-\frac{\sqrt{3}}{2}\right) = -3.5\sqrt{3}.$

17) **Choice D is correct.**

The sum of first 50 natural numbers is the sum of an arithmetic series with the first term $a = 1$ and the common difference $d = 1$. Using the sum formula of arithmetic sequence $S_n = \frac{n}{2}(a_1 + a_n)$, where n is the number of terms, a_1 is the first term, and a_n is the nth term of the sequence. We get:

$$S_{50} = \frac{50}{2}(1 + 50) = 25 \times 51 = 1,275$$

Calculus Practice Tests 2 Explanations

18) Choice A is correct.

$$\frac{1}{2}n(2a + (n-1)d) = \frac{1}{2} \times 20(6 + (19 \times 4)) = 10 \times 82 = 820$$

19) Choice C is correct

Sequence is geometric, $a = 1$ and $r = \frac{1}{2}$, so:

$$s_n = \sum_{i=1}^{n} ar^{i-1} = a\left(\frac{r^n - 1}{r - 1}\right) = 1\left(\frac{\left(\frac{1}{2}\right)^7 - 1}{\frac{1}{2} - 1}\right) = \frac{-\frac{127}{128}}{-\frac{1}{2}} = \frac{127}{64}$$

20) Choice B is correct.

Kth entry in row n: $_nC_{k-1}$. So, the 7th entry ($K = 7$) on the 10th row ($n = 10$) of pascal's triangle becomes: $_{10}C_6$. Therefore,

$$_{10}C_6 = \frac{10!}{6!\,4!} = \frac{10 \times 9 \times 8 \times 7 \times 6!}{6! \times 4 \times 3 \times 2} = 210$$

21) Choice B is correct.

Midpoint of the interval $\left(-\frac{3}{7}, \frac{2}{9}\right) = \frac{-\frac{3}{7} + \frac{2}{9}}{2} = \frac{\frac{-27+14}{63}}{2} = \frac{\frac{-13}{63}}{2} = -\frac{13}{126}$.

22) Choice A is correct.

As $\frac{1}{x}$ approaches zero, left and right limits are different, besides, they would approach infinity anyway.

23) Choice C is correct.

t would be the expression that remain after removing the discontinuity, meaning the expression that remains after simplification:

$$\frac{x^2 - 1}{x - 1} = \frac{(x - 1)(x + 1)}{x - 1} = x + 1 = 1 + 1 = 2$$

Therefore, the value of t is 2.

24) Choice A is correct.

We know that the integer part of a number is a number equal to or smaller than it: $x - 1 < \lfloor x \rfloor \leq x$. So, for every real number as x:

$$\frac{1}{x} - 1 < \left\lfloor \frac{1}{x} \right\rfloor \leq \frac{1}{x}$$

We multiply the sides of each inequality by x with a same sign. Then:

$$x \cdot \left(\frac{1}{x} - 1 \right) < x \cdot \left\lfloor \frac{1}{x} \right\rfloor \leq x \cdot \frac{1}{x}$$

Next, simplify:

$$1 - x < x \left\lfloor \frac{1}{x} \right\rfloor \leq 1$$

Now, we get the limit of the sides of the inequality:

$$\lim_{x \to 0}(1 - x) < \lim_{x \to 0} x \left\lfloor \frac{1}{x} \right\rfloor \leq \lim_{x \to 0} 1 \Rightarrow 1 - 0 < \lim_{x \to 0} x \left\lfloor \frac{1}{x} \right\rfloor \leq 1$$

Therefore, the limit of $x \left\lfloor \frac{1}{x} \right\rfloor$ as x approaches 0 is 1.

$$\lim_{x \to 0} x \left\lfloor \frac{1}{x} \right\rfloor = 1$$

25) Choice C is correct.

Substituting the values for x values that are slightly greater than 2:

$$\lim_{x \to 2^+} \frac{\lfloor x \rfloor - \lfloor x^2 \rfloor}{\lceil -x \rceil} = \frac{\lfloor 2^+ \rfloor - \lfloor 4^+ \rfloor}{\lceil -2^+ \rceil} = \frac{2 - 5}{-2} = \frac{3}{2}$$

26) Choice D is correct.

Limit is ambiguous: $\frac{\infty}{\infty}$, we need to turn it to $\frac{0}{0}$ first, by taking the reciprocal of the cotangent functions and then use the limit properties of the tangent function:

$$\lim_{x \to 0} \frac{\cot(3x)}{\cot(5x)} = \lim_{x \to 0} \frac{\frac{1}{\cot(5x)}}{\frac{1}{\cot(3x)}} = \lim_{x \to 0} \frac{\tan(5x)}{\tan(3x)}$$

Now, this is a sign that you can use L'Hôpital's Rule. You can take the derivative of the denominator and denominator and then calculate the limit again.

$$\lim_{x \to 0} \frac{\tan(5x)}{\tan(3x)} = \lim_{x \to 0} \frac{(5x)'(\tan(5x))'}{(3x)'(\tan(3x))'} = \lim_{x \to 0} \frac{5 \cdot \sec^2(5x)}{3 \cdot \sec^2(5x)}$$

$$= \lim_{x \to 0} \frac{5 \cdot \cos^2(3x)}{3 \cdot \cos^2(5x)} = \frac{5 \cdot \cos^2(3 \times 0)}{3 \cdot \cos^2(5 \times 0)} = \frac{5}{3}$$

27) Choice D is correct.

At $x = 2$, we encounter $\frac{0}{0}$, so we need to simplify the expressions in numerator and denominator to eliminate the part that's causing $\frac{0}{0}$, then finding the limit at $x = 2$: $\lim_{x \to 2} \frac{x^2+2x-8}{x^2-3x+2} = \lim_{x \to 2} \frac{(x-2)(x+4)}{(x-1)(x-2)} = \lim_{x \to 2} \frac{x+4}{x-1} = \frac{2+4}{2-1} = \frac{6}{1} = 6$.

28) Choice C is correct.

Using the power rule of derivatives: $(-2x^3)' = -2 \times 3 \times x^2 = -6x^2$.

29) Choice B is correct.

Instead of using the quotient rule on the original expression, we simplify it to $\left(\frac{3}{-4x}\right)'$ and bring the x from the denominator to numerator by giving it a -1 exponent: $\left(-\frac{3}{4}x^{-1}\right)' = -\frac{3}{4} \times -1 \times x^{-2} = \frac{3}{4x^2}$.

30) Choice B is correct.

To find the derivative, using chain rule, we need to multiply the derivative of $cos(\pi x)$ by the derivative of (πx): $(cos(\pi x))' = -\sin \pi x \times \pi = -\pi \sin \pi x$.

31) Choice A is correct.

Using the chain rule: $\frac{df}{dx} = \frac{dg}{dh} \times \frac{dh}{di} \times \frac{di}{dx} \Rightarrow 7 \times 4i \times -2x = -56i \cdot x$

And since $i(x) = -x^2$: $-56i \cdot x = -56(-x^2) \times x = 56x^3$.

32) Choice D is correct.

Using chain rule and power rule, we have: $3((3x^2 - 2x)^2 \cdot (6x - 2)) = 3((9x^4 - 12x^3 + 4x^2)(6x - 2)) = 3(54x^5 - 18x^4 - 72x^4 + 24x^3 + 24x^3 - 8x^2) = 3(54x^5 - 90x^4 + 48x^3 - 8x^2) = 162x^5 - 270x^4 + 144x^3 - 24x^2$

33) Choice C is correct.

We can use the power rule in combination with chain rule, or we can use the formula $\left(\sqrt{f(x)}\right)' = \frac{f'(x)}{2\sqrt{f(x)}}$:

$$\left(\sqrt{(-2x + 5)}\right)' = \frac{(-2x + 5)'}{2\sqrt{-2x + 5}} = \frac{-2}{2\sqrt{-2x + 5}} = \frac{-1}{\sqrt{-2x + 5}}$$

34) Choice B is correct.

Since we need to use the chain rule, let's find the derivative of the inner function first, then use the $\left(\sqrt{f(x)}\right)'$ formula: $f'(x) = (cos(2x) - 4)' = -2\sin(2x)$.

$$\left(\sqrt{f(x)}\right)' = \frac{f'(x)}{2\sqrt{f(x)}} = \frac{-2\sin(2x)}{2\sqrt{cos(2x) - 4}} = \frac{-\sin(2x)}{\sqrt{cos(2x) - 4}}$$

35) Choice B is correct.

Using $(\log_a f(x))' = \left[\frac{f'(x)}{f(x) \ln a}\right]$:

$$(\log_3 3x)' = \frac{3}{3x \ln 3} = \frac{1}{x \ln 3}$$

36) **Choice A is correct.**

First, finding the first derivative: $(9x^3 - 4x)' = 27x^2 - 4$.

Then, taking the derivative of the first derivative:

$$(9x^3 - 4x)'' = (27x^2 - 4)' = 54x$$

37) **Choice A is correct.**

Finding the first derivative: $(\cos(x) + 3x^2)' = -\sin(x) + 6x$.

Second derivative: $(\cos(x) + 3x^2)'' = (-\sin(x) + 6x)' = -\cos(x) + 6$.

Third derivative: $(\cos(x) + 3x^2)''' = (-\cos(x) + 6)' = \sin(x)$.

38) **Choice B is correct.**

If we take the derivative of both sides with respect to x:

$$x^2 + y^2 = 25 \Rightarrow 2x + 2y \frac{dy}{dx} = 0$$

So, $2y \frac{dy}{dx} = -2x$, therefore: $\frac{dy}{dx} = \frac{-2x}{2y} = -\frac{x}{y}$.

39) **Choice A is correct.**

$$x^2 y = 6 \Rightarrow y = \frac{6}{x^2}$$

Differentiating both sides with respect to x:

$$\frac{dy}{dx} = 6 \times (-2) \times x^{-3} \Rightarrow \frac{dy}{dx} = -\frac{12}{x^3}$$

40) **Choice D is correct.**

$$\int 36x + 6 \, dx = \frac{36x^2}{2} + 6x + C = 18x^2 + 6x + C$$

Because both -1 and -3 could be the constant of integration, choice D is correct.

41) **Choice D is correct.**

$$\int 6x^3\, dx = \frac{6x^4}{4} + C$$

Now using the fundamental theorem of calculus:

$$F(4) - F(1) = \frac{6(4)^4}{4} - \frac{6(1)^4}{4} = 6 \times 4^3 - 1.5 = 382.5$$

42) **Choice C is correct.**

By finding the derivative of a function and setting it to zero, we can find the possible minimum and maximum values of that function in an optimization problem.

43) **Choice A is correct.**

For the interval [1,3], using two subintervals we end up with: (1,2) and (2,3).

The midpoints for these subintervals are at $x = 1.5$ and $x = 2.5$ respectively.

$$\Delta x = \frac{3-1}{2} = 1$$

The midpoint Riemann sum is found using $f(1.5)(\Delta x) + f(2.5)(\Delta x)$:

$$f(1.5)(1) + f(2.5)(1) = 1.5^2 + 2.5^2 = 2.25 + 6.25 = 8.5$$

44) **Choice B is correct.**

We find r by dividing the two terms:

$$\frac{a_5}{a_4} = \frac{ar^4}{ar^3} = r \Rightarrow r = \frac{4.5}{\frac{-27}{2}} = \frac{9}{-27} = -\frac{1}{3}$$

If $a_5 = ar^4 = 4.5$, then we find a:

$$a = \frac{a_5}{r^4} \Rightarrow a = \frac{4.5}{\left(-\frac{1}{3}\right)^4} = \frac{4.5}{\frac{1}{81}} = 364.5$$

So, $S_n = a\left(\frac{1-r^n}{1-r}\right)$:

$$S_3 = 364.5\left(\frac{1-\left(-\frac{1}{3}\right)^3}{1-\left(-\frac{1}{3}\right)}\right) = 364.5\left(\frac{1+\frac{1}{27}}{1+\frac{1}{3}}\right) = 364.5\left(\frac{\frac{28}{27}}{\frac{4}{3}}\right) = 283.5$$

45) Choice C is Correct

Since the absolute value of the ratio is $\frac{2}{3}$ and less than 1, the sum of this infinite geometric series is a finite value. Therefore, by using this formula:

$$\sum_{i=0}^{\infty} a_i r^i = \frac{a_1}{1-r} \Rightarrow \sum_{i=1}^{\infty} \left(-\frac{2}{3}\right)^{i-1} = \frac{1}{1-\left(-\frac{2}{3}\right)} = \frac{1}{\frac{5}{3}} = \frac{3}{5}$$

46) Choice C is correct.

$$\int \sin^2(x)\,dx = \int \frac{(1-\cos(2x))}{2}\,dx = \int \left(\frac{1}{2} - \frac{\cos(2x)}{2}\right)dx$$

We can split this integral into two separate integrals:

$$\int \frac{1}{2}\,dx - \int \frac{\cos(2x)}{2}\,dx$$

The first integral becomes: $\int \frac{1}{2}\,dx = \frac{1}{2}\int dx = \frac{1}{2}x + C_1$.

The second integral can be solved using the trigonometric identity for integrals. Using the trigonometric identity: $\int \cos(ax)\,dx = \frac{1}{a}\sin(ax) + C$.

$$\int \frac{\cos(2x)}{2}\,dx = \frac{1}{2}\int \cos(2x)\,dx = \frac{1}{2}\left(\frac{1}{2}\sin(2x)\right) + C_2 = \frac{1}{4}\sin(2x) + C_2$$

Therefore, the complete solution to the original integral is:

$$\int \sin^2(x)\,dx = \left(\frac{1}{2}x + C_1\right) - \left(\frac{1}{4}\sin(2x) + C_2\right) = \frac{1}{2}x - \frac{1}{4}\sin(2x) + C$$

Where $C_1 - C_2 = C$.

47) **Choice D is correct.**

Using integration by parts, let $u = x^2 \Rightarrow du = 2x\,dx$ and $dv = \cos(x)\,dx$.

After integrating dv: $v = \sin(x)$. After applying the formula, we have:

$$\int x^2 \cos(x)\,dx = x^2 \sin(x) - \int \sin(x) \times 2x\,dx$$

We can do the integration by parts for $\int 2x \sin(x)\,dx$ again. The new u and dv can be: $u = 2x \Rightarrow du = 2\,dx$, and $dv = \sin(x)\,dx \Rightarrow v = -\cos(x)$. So, if we apply the formula of integration by parts again:

$$\int 2x \sin(x)\,dx = -2x \cos(x) - \int (-\cos(x)) \times 2\,dx$$

$$= -2x \cos(x) + 2 \sin(x) + C_1$$

Back to the original problem: $\int x^2 \cos(x)\,dx = x^2 \sin(x) - \int 2x \sin(x)\,dx$:

$$\int x^2 \cos(x)\,dx = x^2 \sin(x) - (-2x \cos(x) + 2 \sin(x) + C_1)$$

$$= x^2 \sin(x) + 2x \cos(x) - 2 \sin(x) + C$$

48) **Choice A is correct.**

Using $\int \cos(nx)\,dx = \frac{1}{n}\sin(nx) + C$, we get: $\int \cos(3x)\,dx = \frac{1}{3}\sin(3x) + C$.

49) **Choice C is correct.**

Using $\int \sqrt{nx}\,dx = \frac{2\sqrt{n}}{3}\sqrt{x^3} + C$, we have: $\int \sqrt{5x}\,dx = \frac{2\sqrt{5}}{3}\sqrt{x^3} + C$.

50) **Choice C is correct.**

We need to find the integral first, then find the equation for definite integral of the function between 1 and b using fundamental theorem of calculus, then take its limit as b approaches infinity:

$\int \frac{1}{x^3} dx = \int x^{-3} dx = -\frac{1}{2x^2}$, and so $\int_1^b \frac{1}{x^3} dx = \left[-\frac{1}{2b^2} - \left(-\frac{1}{2(1)^2}\right)\right] = -\frac{1}{2b^2} + \frac{1}{2}$.

Now, to get the answer, we get the following limit.

$$\lim_{b \to \infty} -\frac{1}{2b^2} + \frac{1}{2} = \frac{1}{2}$$

Because as b approaches infinity, $-\frac{1}{2b^2}$ approaches zero, leaving us with $\frac{1}{2}$.

51) **Choice D is correct.**

Order refers to highest derivative present, and degree is the exponent of this highest derivative. Option D has a 2nd order term, raised to the power of 3, meaning of 3rd degree.

Option A has a 2nd order term, but the first order term is raised to the power of 3.

Option B has a 2nd order term, but it is raised to the power of 2 and the first order term is raised to the power of 3.

Option C has a 3rd order term, raised to the power of 2, meaning the equation is 2nd of degree.

52) **Choice C is correct.**

To be homogenous, all terms must be multiplied by the dependent variable and no free term must exist. This disqualifies option D since it has $sin(x)$.

To be linear, there must be no independent variable terms multiplied by each other or raised to power. So, option B is not the answer, because of $y\frac{dy}{dx}$. Option A is second order and first degree, so option C is the answer.

Please note that $\frac{d^2y}{dx^2}$ indicates second derivative (second order), but $\left(\frac{dy}{dx}\right)^2$

indicates a first order term that is raised to the power of 2.

53) **Choice B is correct.**

We need to solve the equations by separating and integrating them, then we'll see which one matches the curve:

A. $\frac{dy}{dx} = 3x \Rightarrow dy = 3x\,dx \Rightarrow y = \frac{3x^2}{2} + C$

This one is quadratic but the vertex is on the origin and C value only makes the curve to move up and down, not left and right, so this option can't be the answer.

B. $\frac{dy}{dx} = 3x - 5 \Rightarrow dy = 3x - 5 \Rightarrow y = \frac{3x^2}{2} - 5x + C$

This option matches the curve provided in the problem.

C. $\frac{dy}{dx} = -x \Rightarrow dy = -x\,dx \Rightarrow y = -\frac{x^2}{2} + C$

The curve of this option is flipped horizontally, so this can't be the answer.

D. $\frac{dy}{dx} = -3x \Rightarrow dy = -3x\,dx \Rightarrow y = -\frac{3x^2}{2} + C$

Which is also flipped with respect to $x-$axis, and can't be the answer.

54) **Choice A is correct.**

The differential equation describing the decay of Nobelium, which has a half-life of 58 minutes is given by $\frac{dN}{dt} = -\frac{\ln 2}{58} \times N(t)$, with $\frac{dN}{dt}$ representing the rate of change of Nobelium's quantity over time, $N(t)$ being the amount remaining at time t, and $\frac{\ln 2}{58}$ is the decay constant derived from its half-life, with the minus sign indicating the loss.

55) **Choice A is correct.**

Using $x = r\cos(\theta)$ and $y = r\sin(\theta)$. For $r = 5$ and $\theta = \frac{\pi}{3}$: $x = 5\cos\left(\frac{\pi}{3}\right) = \frac{5}{2}$ and $y = 5\sin\left(\frac{\pi}{3}\right) = \frac{5\sqrt{3}}{2}$. So, the coordinate will be $\left(\frac{5}{2}, \frac{5\sqrt{3}}{2}\right)$.

Calculus Practice Tests 2 Explanations

56) Choice A is correct.

$(3 + 9i)(-2 - 6i) = -6 - 18i - 18i - 54i^2$

$= -6 - 36i - 54(-1) = -6 + 54 - 36i = 48 - 36i$

57) Choice C is correct.

$$\frac{4 + 3i}{7 - i} \times \frac{7 + i}{7 + i} = \frac{28 + 4i + 21i + 3i^2}{49 + 7i - 7i - i^2} = \frac{28 + 25i + 3(-1)}{49 - (-1)}$$

$$= \frac{28 - 3 + 25i}{50} = \frac{25(i + 1)}{50} = \frac{i + 1}{2}$$

58) Choice B is correct.

Using quotient rule of derivatives: $\left(\frac{sin(3x)}{\sqrt{x^3}}\right)' = \frac{(sin(3x))' \cdot \sqrt{x^3} - \left((\sqrt{x^3})' \cdot (sin(3x))\right)}{(\sqrt{x^3})^2}$.

Using the power rule and chain rule: $\left(\frac{sin(3x)}{\sqrt{x^3}}\right)' = \frac{3\cos(3x) \cdot \sqrt{x^3} - \frac{3}{2}\sqrt{x} \sin(3x)}{x^3}$.

59) Choice C is correct.

The formula represents the Binomial Theorem, used for expanding expressions raised to a power and calculating probabilities in binomial distributions in statistics.

60) Choice B is correct.

Integrals, the inverse of derivatives, compute total accumulations, like areas under curves or accumulated quantities over time.

Effortless Math's Calculus Online Center

... So Much More Online!

Effortless Math Online Calculus Center offers a complete study program, including the following:

- ✓ Step-by-step instructions on how to prepare for the Calculus test

- ✓ Numerous Calculus worksheets to help you measure your math skills

- ✓ Complete list of Calculus formulas

- ✓ Video lessons for Calculus topics

- ✓ Full-length Calculus practice tests

- ✓ And much more…

No Registration Required.

Visit **EffortlessMath.com/calculus** to find your online Calculus resources.

Build Your Math Skills: Our Top Book Picks!

Download eBooks (in PDF format) Instantly!

Download

Our Most Popular Books!

 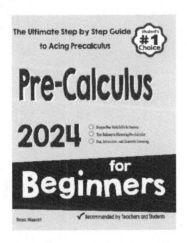 

Our Most Popular Books!

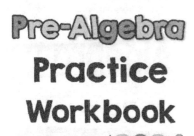

Receive the PDF version of this book or get another FREE book!

Thank you for using our Book!

Do you LOVE this book?

Then, you can get the PDF version of this book or another book absolutely FREE!

Please email us at:

info@EffortlessMath.com

for details.

Author's Final Note

I hope you enjoyed reading this book. You've made it through the book! Great job!

First of all, thank you for purchasing this study guide. I know you could have picked any number of books to help you prepare for your Calculus course, but you picked this book and for that I am extremely grateful.

It took me years to write this study guide for the Calculus because I wanted to prepare a comprehensive Calculus study guide to help students make the most effective use of their valuable time while preparing for the final test.

After teaching and tutoring math courses for over a decade, I've gathered my personal notes and lessons to develop this study guide. It is my greatest hope that the lessons in this book could help you prepare for your test successfully.

If you have any questions, please contact me at reza@effortlessmath.com and I will be glad to assist. Your feedback will help me to greatly improve the quality of my books in the future and make this book even better. Furthermore, I expect that I have made a few minor errors somewhere in this study guide. If you think this to be the case, please let me know so I can fix the issue as soon as possible.

If you enjoyed this book and found some benefit in reading this, I'd like to hear from you and hope that you could take a quick minute to post a review on the book's Amazon page.

I personally go over every single review, to make sure my books really are reaching out and helping students and test takers. Please help me help Calculus students, by leaving a review!

I wish you all the best in your future success!

Reza Nazari

Math teacher and author

Made in United States
North Haven, CT
01 April 2024